Cerebrum 2007

Cerebrum 2007

EMERGING IDEAS IN BRAIN SCIENCE

Cynthia A. Read, Editor

DANA
PRESS

New York • Washington, D.C.

DANA
PRESS

The Dana Foundation
745 Fifth Avenue, Suite 900
New York, NY 10151

900 15th Street NW
Washington, D.C. 20005

DANA is a federally registered trademark.

ISBN 13: 978-1-932594-24-9

ISSN 1524-6205

Cover design by Christopher Tobias Design
Cover illustration by Anna & Elena Balbusso
Text design by Kachergis Book Design

Contents

Book Reviews

Introduction

by Bruce McEwen, Ph.D.

Bruce S. McEwen, Ph.D., is a professor and head of the Laboratory of Neuroendocrinology at The Rockefeller University, where his research focuses on the actions of stress and sex hormones on the brain and immune system. He is author of *The End of Stress as We Know It* (Dana Press/Joseph Henry Press, 2002), and *The Hostage Brain* (The Rockefeller University Press, 1994). He can be reached at mcewen@rockefeller.edu.

WHETHER WE ARE TALKING ABOUT terrorism, finding the collective will to confront global climate change, or confronting the causes and possible cures for the increasing individual and societal burden of depression, the greatest problems facing humankind these days seem to deal with human behavior and misbehavior. The World Health Organization's Global Burden of Disease report lists seven brain- or behavior-related disorders among the top ten that produce disability and economic burden. For many parents, worries about developmental conditions such as autism and attention deficit disorder are now at the forefront, while at the other end of the life span aging baby boomers fear Alzheimer's disease.

Behavior is a function of the brain. Neuroscience is teaching us that the brain is an organ that is shaped both by genes, starting at conception, and by experiences throughout life. And yet the genetic influence is often subtle, involving normal variants of many genes, and, in most cases, is manifested only as a result of life events. For example, depressive illness is more likely in people who have a common variant of the molecule that transports the important brain chemical serotonin combined with adverse life experiences.

Brain science is helping us understand the biological underpinnings of fear, love, hate, and aggression, as well as arousal, decision making, reward, and the perturbations of reward in substance abuse. The biological basis of memory is within our conceptual and experimental grasp. The consequences for the brain of stress and malnutrition, on the negative side, and exercise and environmental enrichment, on the positive side, are now recognized in increasing detail and offer possibilities for treating disorders from depression and schizophrenia to autism, obsessive compulsive disorder (OCD), and substance abuse.

Moreover, the brain regulates the health of the body through the autonomic nervous system and through hormones and the immune system. Thus conditions that we don't commonly consider brain disorders, like obesity, cardiovascular disease, arthritis, autoimmune disorders, and chronic pain, are problems in neuroscience because they also affect brain function and are influenced by behavior.

Scientists are often accused of being in ivory towers and not dealing with real-world problems, but *Cerebrum: The Dana Forum on Brain Science* gives voice to the thoughtful integration of ideas by experts in various fields. This endeavor has great value at a time of increasing specialization of technology and information into silos of knowledge, because disciplinary journals do not allow us to see that knowledge in a big picture across disciplines. *Cerebrum 2007* brings together more than a dozen articles and book reviews from the journal's Web edition to let book readers be among the first to learn from top thinkers in their fields what may well become tomorrow's conventional wisdom on topics such as the biological nature of ethical behavior, the brain basis for belief in the supernatural, the science of music, and the use of medications to alter traumatic memories. The articles in *Cerebrum 2007* discuss some of the conditions that can hold a person's brain hostage, such as autism, as well as OCD, chronic pain, sleep disorders, post-traumatic stress disorder, and the terrible consequences of traumatic brain injury. As in many other aspects of clinical neuroscience, including psychiatry, brain imaging has been an invaluable tool.

Autism is a spectrum of developmental disorders for which there is no apparent single cause and many behavioral manifestations, according to Diane Chugani and Kayt Sukel in "Bringing the Brain of the Child with Autism Back on Track." Yet considerable hope lies in the use of behavioral training as well as management of symptoms with pharmaceutical agents. OCD is another disorder with multiple aspects that is now recognized as an organic brain disorder. It affects both anxiety and higher cognitive function in the frontal lobes, write Judith Rapoport and Gale Inoff-Germain in "Unshackling the Slaves of Obsession and Compulsion: A Brain Science Success Story." OCD can be treated with cognitive behavioral therapy, along with medications such as the serotonin reuptake inhibitors, also used to treat depression. Another huge problem that dominates the lives of many people is inadequate treatment of pain for those suffering from fibromyalgia and neuropathic pain disorders, as well as cancer. In "Why Not a National Institute on Pain Research?" Kathleen Foley and Maia Szalavitz address the public perceptions and medical reality of the use of opioids as pain relievers and the degree to which addiction should or should not be a major concern.

"Toward a New Treatment for Traumatic Memories" is the subject of an article by Margaret Altemus and Jacek Dębiec. Since 9/11, and also because of the Vietnam, Gulf, and Iraq wars, concern has increased about post-traumatic stress disorder in veterans, as well as among civilians exposed to horrible events, and how, if possible, to suppress or rid the brain of memories of violence. The new view of such memories is that they are labile; we might be able to reactivate and then suppress them by the right pharmacologic manipulations. In addition, we now know that a fear-related memory can be overridden by new information that the danger is no longer present. This is one of the goals of cognitive behavioral therapy. However, drugs that might help the brain get rid of traumatic memories could also be misused, and this topic raises ethical concerns that must be dealt with.

The Terry Schiavo case revealed how ignorance can distort ethical concerns and help to politicize a terrible human tragedy. In a classic article from 2003—an instance of how *Cerebrum* often is ahead of the news—Nicholas Schiff and Joseph Fins define the three states resulting from traumatic brain injury that are often confused with each other: persistent vegetative state (which afflicted Terry Schiavo), coma, and minimally conscious state. They discuss the ethical issue surrounding the termination of life in a person in a persistent vegetative state and also the possibilities, as well as the ethical dilemma, associated with efforts to rouse the brain through deep brain stimulation (DBS). Only recently have major news organizations widely reported on brain scientists' discoveries about the various "comatose" states and on the potential of DBS, primarily when used for Parkinson's disease and depression. The possibility of using DBS for people who have experienced a traumatic brain injury, as suggested by Schiff and Fins in "Hope for 'Comatose' Patients," is still relatively unknown.

Remarkable progress in the neuroscience of how the brain responds to damage by means of its own plasticity is leading to treatments that take advantage of and accentuate the ability of the brain to repair itself after stroke, according to "Searching for a New Strategy to Protect the Brain" by Nicolas Bazan. Yet, as discussed by Louis Caplan in "Improving Stroke Prevention and Treatment Now," existing and hoped-for treatments for stroke and related cardiovascular disorders are the last line of defense and must be considered together with recognizing and, where possible, altering the factors in experience and behavior that make some people more vulnerable than others. These include both genetic traits and environmental causes such as those factors that lead to obesity, diabetes, and cardiovascular disease—lack of physical activity, overuse of alcohol, smoking, and poor diet. Economically disadvantaged and less-well-educated people are among those most frequently and seriously affected by these diseases, for a variety of reasons.

Often overlooked in the discussion of vulnerability to anxiety and mood disorders, as well as difficulties in coping with daily life and its consequences for physical health, is the matter of individual temperament. In "Hardwired for Happiness," behavioral biologist Silvia Cardoso discusses happiness as a trait that is part of a larger spectrum of positive emotions, including optimism and exuberance. A better understanding of happiness and its biological underpinnings may be helpful in matching people to careers and in dealing with mental health problems, for example feelings of helplessness and low self-esteem. Moreover, there are indications that a positive outlook on life is associated with lower vulnerability to stress, a discovery that reinforces our growing appreciation and better understanding of the mind-body interconnection.

While much can be done through behavior and lifestyle change to slow down or prevent certain diseases, the development of new pharmaceutical agents is essential for treatment of many disorders, from Alzheimer's disease to schizophrenia to addiction. Paul M. Matthews discusses in his article "Brain Imaging for Better Drug Development" how brain imaging using PET and fMRI can speed up the current dauntingly long process of developing new medications.

Modern imaging techniques, together with recording of electrical activity from the brain, have have also provided important clues to help explain why some people become sleepwalkers, according to an article on "Are We in the Dark About Sleepwalking's Dangers?" by Shelley R. Gunn and W. Stewart Gunn. Sleep deprivation, especially in combination with drugs and alcohol, is known to induce sleepwalking in some people, and the resulting behavior is extremely unpredictable, particularly in a new environment. Continued deactivation of the prefrontal cortex of the brain, in combination with abnormal activation of the cingulate cortex and the thalamus, may lead to the dissociation between "body sleep" and "mind sleep" characteristic of somnambulism.

Besides its vulnerability as a biological organ, the human brain is amazing and fascinating for its ability to learn and remember, to love and hate, to create great art and music, and to strive to understand itself and the universe. David Huron reviews a new book on the science of music by Daniel Levitin, *This Is Your Brain on Music: The Science of a Human Obsession*, in which the neurobiological mechanisms of how the nervous system responds to the musical experience are explored. Kevin Tracey reviews a book by Ruth McKernan, *Billy's Halo: Love, Science and My*

Father's Death, which deals with grief, love, and pursuit of understanding of how and why a man died from a septic infection. And Edward McKintosh reviews a book on the history, promise, and perils of neural prosthetic devices by Victor Chase, *Shattered Nerves: How Science is Solving Modern Medicine's Most Perplexing Problem.*

When we speak of the "mind," it is clear that we mean something more than the brain as an organ. According to the article by Bruce Hood, "The Intuitive Magician: Why Belief in the Supernatural Persists," the mind has an intrinsic need to fill in the gaps. When there is no scientific explanation, it attempts to make sense of the world by creating explanations that sometimes include supernatural beings and processes. In this light, the recent resurgence of interest in intelligent design can be regarded as a manifestation of the need to understand how human beings and other forms of life appeared on this planet.

Speaking of the tendency of the mind to impute meaning in nature, in their article "Elephants that Paint, Birds That Make Music: Do Animals Have an Aesthetic Sense?" Lesley Rogers and Gisela Kaplan discuss both sides of the question of whether animals create and enjoy what human beings call art and music. On the one hand, because of our own aesthetic sense we may interpret as art paintings by elephants or monkeys or delightful bird songs. On the other hand, in the face of growing evidence for animals' complex cognitive abilities, the authors suggest that we should not be too hasty in deciding whether what is art to us might also be art to them.

The human mind also attempts to distinguish right from wrong, and in "A Brain Built for Fair Play" Donald Pfaff speculates that fear and anxiety, coupled with empathy and a conscious awareness of how we ourselves would like to be treated by others, drives a sense of fair play in most human beings. The discovery of mirror neurons—neurons that respond when an observer sees another individual doing something—has provided a potential basis for understanding the process of empathy in neurobiological terms. Yet the advances in neuroscience also highlight past mistakes, such as those resulting from prefrontal lobotomy and other types of psychosurgery, as discussed by Henry Greely in "Knowing

Sin: Making Sure Good Science Doesn't Go Bad." He also notes that new ethical challenges come from the possible use of brain imaging to detect states of people's minds, as well as from the use of performance-enhancing drugs. In addition, the sequencing of the human genome and recognition of genetic contributions to many disorders has once again raised the specter of eugenics and of attempts to engineer the most desirable traits and eliminate undesirable ones. Related to this is the debate over when life begins during development and the ethics of terminating pregnancy. The formation of a new neuroethics society, announced in *Cerebrum*, is a welcome event given these growing concerns.

This brings us back to asking how much the external world that we all experience every day affects the internal world of the brain and shapes it for better or worse. Does the human mind have an intrinsic striving for goodness and cooperation, or are hatreds based on race and religion, or the consequences of maltreatment, or the competition for resources so powerful that they overcome these basic tendencies, if they exist? Can we learn to work together to solve problems common to all of us? No time is more important than the present for the brain sciences to join in this discussion, since the survival of life on our planet as we know it may well depend on our ability to put aside differences and work together for the common good. Or, as the Dutch, most of whom live below sea level, say: "We all live on the same polder."

CHAPTER I

(TWO ARTICLES)

Stroke *We Can and Must Do Better*

Louis R. Caplan, M.D. *and*
Nicolas G. Bazan, M.D., Ph.D.

Louis R. Caplan, M.D., is a professor of neurology at Harvard Medical School and chief of the Cerebrovascular/ Stroke Division at Beth Israel Deaconess Medical Center. He is the founder of the Harvard Stroke Registry and the author of numerous articles and books about stroke, including *Striking Back at Stroke: A Doctor-Patient Journal* (Dana Press, 2003) and *Stroke* (Demos Medical Publishing, 2005). He can be reached at caplan@bidmc.harvard.edu.

Nicolas Bazan, M.D., Ph.D., is a professor of ophthalmology, biochemistry and molecular biology, neurology, and neuroscience at Louisiana State University Health Sciences Center, and founding director of LSU's Neuroscience Center of Excellence, where research is moving forward despite the post-Katrina challenges. Through his research on synaptic signaling pathways he hopes to develop ways to reduce or prevent the irreversible brain damage caused by stroke, epilepsy, and neurodegenerative disease, as well as retinitis pigmentosa and age-related macular degeneration. He can be reached at NBazan@lsuhsc.edu.

Improving Stroke Prevention and Treatment Now

Louis R. Caplan, M.D.

DURING THE PAST TWENTY-FIVE YEARS, more advances were made than ever before in our understanding of strokes and in our ability to prevent and treat them. Technology is readily available that can quickly and safely image the brain and heart and the blood vessels that supply them. Drugs can effectively treat and control risk factors that lead to stroke. Other drugs, surgery, and other interventions can now minimize stroke-related brain damage. Yet strokes continue to happen at an alarming rate, and stroke continues to be the third leading cause of death in the world and probably the most important cause of long-term morbidity. When people are asked to share their worries about their health as they age, they invariably mention two concerns: cancer and the pain related to it, and becoming disabled and dependent—losing the ability to communicate, think, use their arms and legs normally, or walk, all of which can be the result of a stroke.

Why is the medical profession not doing better at caring for people with a potential for stroke and those who have had a stroke? Why does the disconnect occur between what can be done and what is being done? We must do better.

Controlling Risk Factors for Stroke

Preventive strategies emphasize controlling risk factors that lead to stroke. Of course, some of the risks are beyond a person's control. Men, older people, and those with strong family histories of hypertension, heart disease, and strokes have a higher risk of stroke than people without these histories. But people cannot choose their parents or sex at birth, and all strive to become seniors one day. Many risk factors, however, can be controlled.

Chief among them are:

- Hypertension (high blood pressure)
- Diabetes
- Smoking
- High cholesterol
- Obesity
- Inactivity, lack of adequate exercise
- Recreational drug use, especially cocaine and amphetamines
- Atrial fibrillation (rapid irregular twitching of muscle in upper chamber of the heart)
- Overuse of alcohol

Studies have estimated the number of strokes attributable to some of these risk factors and how many strokes could be prevented by their effective management.[1,2] The results, shown in Figure 1, are dramatic.

Hypertension is clearly the most important risk. Being overweight, a serious problem worldwide, increases the frequency of hypertension.

Figure 1   Where Should We Focus Our Efforts?

Number of Preventable Strokes, based on estimated 700,000 annual strokes

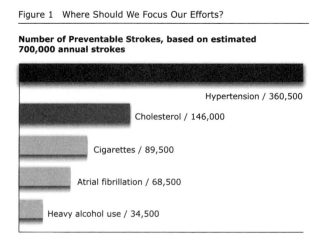

Hypertension / 360,500

Cholesterol / 146,000

Cigarettes / 89,500

Atrial fibrillation / 68,500

Heavy alcohol use / 34,500

Unfortunately, blood pressure control is far from adequate in the United States, as you can see in Figure 2.[3]

Figure 2   Blood Pressure Control in the U.S

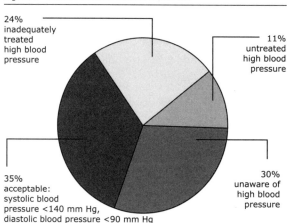

24%
inadequately
treated
high blood
pressure

11%
untreated
high blood
pressure

35%
acceptable:
systolic blood
pressure <140 mm Hg,
diastolic blood pressure <90 mm Hg

30%
unaware of
high blood
pressure

   Clearly, many people are unaware of having high blood pressure. Even when the diagnosis of hypertension is made, management is not as good as it could be. Why? Some explanations relate to patients, some to doctors, and some to medical care systems. People are accustomed to consulting doctors only when they become sick and to taking pills when they feel ill and stopping them when they feel better. Chronic conditions are more difficult to manage than acute illnesses. Hypertension, at least initially, may be accompanied by no important symptoms or dysfunction. Often the prescribed pills have side effects—fatigue, light-headedness, depression, impaired sexual function—so that people may feel better when not taking them. In addition, doctors' practices are customarily not focused on prevention ("well-adult" care), and remuneration for this type of care is scanty.

   Prevention also means systems need to be in place to issue reminders to both doctors and patients about follow-up. Nurses and other medical staff in addition to physicians could perform some of the monitoring and care. The practice of measuring blood pressure with wearable twenty-four-hour monitors is not used as widely as it should be.

A casual, infrequent measurement of blood pressure in a doctor's office may not reflect accurately a person's blood pressure levels during activities of daily living, stress, and sleep.

Many of the other risk factors for stroke—diabetes, obesity, smoking, overuse of alcohol, and use of recreational drugs—are huge problems both in the United States and throughout the world. Management of these factors requires both sustained willpower and support from family, friends, and the medical care system.

Effective agents used to prevent stroke include statins, angiotensin-converting enzyme (ACE) inhibitors, ACE receptor blockers, agents that decrease platelet functions, and anticoagulants. Statins help lower cholesterol but also reduce plaque formation within arteries. Unfortunately, many doctors still only use cholesterol levels as an indication to prescribe statins, and often too low a dose is prescribed. Much evidence now shows that higher doses (the equivalent of 40–80 mg atorvastatin) are more effective than lower doses. ACE inhibitors have salutary effects on vascular endothelia (the internal lining of blood vessels). Platelet inhibitors are also effective in preventing stroke but are often either not prescribed by physicians or not taken by the patient. Anticoagulants in the form of warfarin compounds definitively prevent stroke in patients with atrial fibrillation. Newer anticoagulants in the form of direct thrombin inhibitors are now being tested that may be more effective than heparin and warfarin and cause less bleeding. Often these platelet antiaggregants and anticoagulants are discontinued before minor procedures (for example, dental work, colonoscopy, or minor skin surgery) to avoid bleeding, but strokes and heart attacks occur while the patients are off these drugs, one of the still-unresolved dilemmas of prevention. In addition, many of these medications are expensive and beyond the economic capabilities of all too many older patients.

Starting Young

I must emphasize another key problem. Preventive measures need to be started much earlier in life than they now are—in fact, during childhood. A presentation I heard greatly impressed me. A schoolteacher in the southern United States described an experiment she performed with her sixth grade class. She asked the children, a mixed racial group of twelve- and thirteen-year-olds, to list medical conditions prevalent in their families. Few of the children could. Parents are reluctant to tell young children about their illnesses for fear of frightening them. The teacher then gave the children the assignment to inquire about illnesses and conditions in their family: parents, grandparents, aunts, uncles, and other close relatives. The children were also examined by a pediatrician, and blood tests were performed. Children whose parents had hypertension had higher than normal blood pressures. Children whose families were obese were overweight. In families with a high frequency of diabetes, children's blood glucose was often high. All these conditions often begin quite early in life. After I heard this presentation, I decided to check the cholesterol levels of my six children. I had recently discovered that my cholesterol level was high, and five of my six children had cholesterol levels that were abnormal for their ages. My wife and I then began strategies to lower our cholesterol and fat intakes and those of our children.

When patients are hospitalized with a stroke or a heart attack, doctors often counsel them to change the amount and type of foods they consume and to get more exercise. But can the habits and customs of seventy- and eighty-year-old people be changed? A more important recommendation would be to urge them to have their children and grandchildren checked and to institute good health practices early in life. The eating and exercise habits of children need improvement. Sitting for hours before the television and chowing down fast-food hamburgers, fries, and sugary soda surely increase the likelihood of a premature heart attack or stroke.

Stroke's Early Warning

Before a stroke, many patients have temporary decreases in blood flow to portions of the brain. These episodes are usually referred to as transient ischemic attacks, or TIAs. The frequency of strokes after TIAs is high, and, critically, the hours, days, and weeks after a TIA carry the highest risk of stroke. But people often do not recognize that a symptom they have could, in fact, be a TIA. If a hand goes numb or weak, they often attribute this to a local problem, for example, pressure on the arm.

TIA episodes warrant urgent evaluation to detect the cause. Some are the result of severe disease of a carotid or vertebral artery in the neck or of a large artery within the head, and others are caused by small emboli (blood clots) from heart conditions and atherosclerosis of the aorta. These causes of stroke are treatable, and the treatment may prevent a stroke. Unfortunately, many TIAs are not only ignored by patients but also inadequately investigated by physicians. Some managed-care insurers and other payers refuse to pay for hospital evaluation of TIAs. A quick and thorough evaluation is essential and should be promoted, not discouraged. Insurers should recognize that strokes are costly, so their prevention saves money as well as the health of the people who might otherwise develop a stroke.

Acute Treatment of Strokes

If prevention measures fail and a stroke occurs, what can and should be done? Newer technology that uses computer-assisted tomography (CT), magnetic resonance imaging (MRI), and ultrasound scanning can now reveal quickly and safely the presence of stroke-related brain damage and the heart, blood, and blood vessel abnormalities that are causing brain dysfunction. Having discovered the nature and extent of the problem, doctors have effective drugs and surgical and interventional techniques, such as angioplasty and stenting (surgical procedures to re-open blocked arteries), that can address the abnormalities. As an example, thrombolytic drugs ("clot-busters") can dissolve clots blocking arteries that supply blood to the brain. These agents were approved for clinical use in 1996, and guidelines for their use were written and disseminated by committees of the American Heart Association[4] and the American Academy of Neurology.[5] Yet ten years later, fewer than 5 percent of people eligible for thrombolytic treatment actually receive it. What explains this and how can it be improved?

Problems in treating acute strokes have various origins: patients and their families, doctors and hospitals, and our current medical care systems.

The Person with a Stroke

All too often, people who have a stroke do not get to medical centers in time for effective acute treatment, frequently because they do not recognize that their symptoms could be stroke-related. The public is not well informed about the brain and its functions in general and, specifically, about stroke. If people on the street are asked, "What happens to someone during a 'brain attack'?" the most common answers are that a person will act crazy, have a seizure, go unconscious, or become dumb. People do not usually think of the brain as the repository of inputs from their vision, hearing, and sensations in their arms and legs and as the organ controlling their movement. When an arm feels numb or does not work as usual, many people attribute the dysfunction to a muscle or nerve within that arm. Visual difficulties are attributed to a problem in the eyes, not the brain.

Some people, of course, do realize that a symptom may be related to the brain and indicate a stroke, but, for a variety of reasons, they deny or choose to ignore the possibility, hoping it will go away. Because the brain is the organ that both recognizes a symptom and responds to that recognition, loss of some brain functions can impair the ability to identify the nature of the symptoms and to respond to them. The brain impairment can also render people unable to act, especially if they are alone. If a person cannot talk and his limbs are paralyzed, he cannot initiate a trip to a medical center by himself. Sometimes, too, the person having the stroke (or a family member or friend who is present) does not drive or does not have a car and waits for someone such as a son or daughter to drive to the hospital. Heavy traffic or long distance from a medical center further delays arrival at the emergency room.

A rapid chain of response must occur in order to optimally manage people having a stroke. The possibility of stroke must be recognized quickly, and the stroke patient or a family member should immediately call 911. Calling a local physician or health maintenance organization only delays care. Then an ambulance, or someone accompanying the person with the stroke, should deliver that person to the hospital best equipped for stroke care. Unfortunately, many communities do not have adequate systems. In nearly every urban area, some medical centers have qualified as trauma centers, and emergency systems are in place to get the injured patients to those centers. The same approach should be instituted for stroke.

Doctors and Hospitals

When acute stroke therapy using thrombolytic drugs was approved by the regulating agencies in 1996, many doctors and medical centers were unprepared to put it into practice. Patients with stroke are first seen and examined in the emergency rooms of hospitals staffed by physicians who specialize in emergency care. But, despite the large number of emergencies with neurological aspects, usually these physicians have little training in neurology or in examining and evaluating patients with strokes or other acute neurological problems. Worse, some emergency physician organizations have issued statements that thrombolytic drugs have not been shown to be standard treatment for acute strokes. In many emergency rooms, the emergency physicians will not give thrombolytics without a neurologist, but neurologists are often not readily available.

Neurologists, on the other hand, are trained and experienced in treating strokes but are usually based in offices, not hospitals. To examine a patient in the emergency room, a neurologist often must leave an office full of patients. In addition, a trip to the hospital to examine and treat a patient with an acute stroke often takes hours and is poorly remunerated. For all of these reasons, many neurologists are also not enthusiastic about giving thrombolytic treatment.

Strokes are complicated. The brain anatomy, the many varied conditions that cause stroke, and the blood vessels that bring blood to and away from the brain are all quite complex. Some neurologists have had specialized training—even fellowships—in stroke, but there are too few stroke neurologists. Clearly, more are needed. Today, many hospitals are staffed by what are called "hospitalists"—doctors (usually internal medicine specialists) who work full time in the hospital to manage acute medical problems. Few hospitalist neurologists exist, but medical centers that recruit neurologists especially trained in acute neurological conditions such as stroke may be the best solution to the manpower problem.

Some hospitals are also unprepared to apply the latest techniques in treating strokes. Brain and vascular imaging methods have improved rapidly. Newer MRI and CT scanners have more capability, but the equipment is costly and continued updating is expensive. As a result, state-of-the-art equipment, and stroke neurologists and neuroradiologists competent to use it, are often miles away from the emergency room where a stroke patient is awaiting care.

In addition to recruiting appropriate specialists and acquiring advanced technology, hospitals must generate systems to ensure rapid evaluation of patients with stroke, efficient shepherding throughout the process (what hospitals call "throughput"), and follow-up. Hospitals often self-designate and advertise their capabilities. To attract patients, hospitals may call themselves "stroke centers" even though they may not have adequate personnel, technology, and systems for effectively managing patients who had a stroke. Consequently, the American Stroke Association and many states have begun to issue requirements for designation of "primary" and "secondary" stroke centers. The difference between the designations is that primary centers have the rudimentary necessities for managing stroke, whereas secondary centers have advanced specialized capabilities for medical, surgical, and interventional treatments.

What is the solution to these hospital issues? I urge that the following steps be taken. Hospitals must acquire the appropriate personnel and technology and put in place the systems needed to qualify as a stroke center, or they should not accept patients who may have had a stroke. Those patients should be sent instead to a nearby qualified stroke center. All hospitals cannot—and should not—specialize in everything. For hospitals that cannot become qualified stroke centers and have no such centers nearby to which they can refer patients, a new alternative is "telemedicine." Computer technology enables doctors at a local hospital to have patients assessed quickly by specialists at stroke centers. Doctors at the stroke center can talk to the patient, watch and direct examinations, and review scans sent by computer, then consult with the local physicians on managing the case. Telemedicine is a growing field and now in use in France, Germany, and parts of the United States. This option seems especially attractive for rural hospitals that are far away from stroke centers.

The Continuing Challenge of Recovery

Even the best prevention and acute treatment will not completely eliminate the occurrence of strokes. As people live longer and acquire more medical problems, strokes will continue to happen. So we must also direct our attention to recovery. What medications and rehabilitation strategies facilitate recovery? Which drugs and procedures delay and impede it? Until recently, little research had been done on how people recover from stroke. We are now trying to find out whether brain regions that were temporarily dysfunctional improve, new areas take over for injured regions, or people learn to do activities differently, using brain regions that were not injured. As more patients survive the acute phase of a stroke, both clinical neurologists and neuroscientists must focus on facilitating recovery. Newer technologies, including functional MRI and magnetic stimulation, are yielding much information on what happens after a stroke and enabling a more scientific evaluation of rehabilitation strategies. Their increased efforts will undoubtedly lead to better rehabilitation strategies and programs soon.

Stroke prevention and treatment have come a long way, but not far enough. Not as far as they could, given what we know now, and certainly not as far as would be ideal. Much can and should be done now to optimize care of this critical public health problem.

Searching for a New Strategy to Protect the Brain

Nicolas Bazan, M.D., Ph.D.

TASHA AWOKE EARLY, unable to feel the right side of her face: it was numb. At first, she thought she was still dreaming, but she heard her husband downstairs in the kitchen, and knew that she was awake. She got herself up and checked the mirror for marks on the side of her face from sleeping in an awkward position. She saw nothing. When she put her hand to her face, she suddenly felt a profound weakness in her right arm. Through habit and an act of sheer will, Tasha washed and dressed, intending to make her way to work, where she served as the executive assistant to a bank vice president. Tasha had a full day ahead of her, and whatever was going on would have to be dealt with later.

A fifty-two-year-old African American woman, Tasha was always there for everybody—her growing children, her husband, her boss, and her co-workers—which did not leave much time for herself. Things such as regular exercise and a healthy diet were luxuries she could not afford. She had steadily gained weight during the past two decades and felt the extra burden of her size as she tried to make her way down the stairs.

At the top of the stairs, she felt her right leg buckle; it also was numb and weak. Vision in Tasha's right eye became blurry, and the straight staircase looked like a spiral. Feeling dizzy, she sat down on the landing; her balance and coordination had left her. She called out in fear, and her husband rushed to her. But when he asked what was wrong, Tasha felt confused. She had the worst headache of her life. Tasha's husband called 911, and she was rushed to the hospital. Tasha had had a stroke.

Stroke—"brain attack"—continues to be a killer, and advances in treatment have been disappointing. Several clues to its causes that have emerged in population studies are not fully understood, and their significance in developing effective treatments is not clear. Although stroke occurs with similar frequency in men and women, women have strokes at a younger age and die more frequently. African Americans are three to five times more likely to have strokes than are Caucasians. Risk factors include high-fat diet, obesity, high blood pressure, diabetes, and smoking. Correcting these habits and conditions should reduce the incidence of stroke, but stroke is complex and represents a major scientific and medical challenge.

A Strategy to Protect the Brain

Developing a safe and effective therapy to protect the brain after a stroke, a process known as "neuroprotection," represents a major unsolved challenge for researchers. Current medications prescribed after stroke do not do this, and a great deal of research has focused on developing a safe and effective therapy for use after stroke. However, while a long list of neuroprotective compounds has shown encouraging results in experimental models in animals, most of these potential medications have resulted in disappointing outcomes after clinical trials in humans.[1-3]

We can identify many reasons for this failure. First and foremost, the onset of stroke triggers a complex process in the brain that includes multiple cellular and molecular events occurring at different times. As a result, the target at which any particular drug is aimed may not be a key factor in disease development. In an ischemic stroke—that is, one caused by a blood clot—the brain is very sensitive to the phase when the blood begins to flow again, what is called "reperfusion." Therefore, the time window during which a neuroprotectant might help is also important. In addition, the method of delivering of the drug to a stroke patient is critical, because a drug's concentration, at a specific cellular site in the brain, during a relatively prolonged period of time, must be appropriate. Moreover, the actual design of the clinical trial itself, including the timing for evaluating the outcomes, affects the results. Another challenge is that animal models of a disease such as stroke often show only part of the clinical picture seen in humans. Although animal experiments will continue to contribute new knowledge and significant breakthroughs, we must be careful about the design of those studies and not overinterpret the results.

Many efforts have been made to block the destructive processes that stroke triggers in the brain, and often emphasis is placed on understanding how brain cells die. In my laboratory, we have used a different approach: we try to gain insight into how the brain defends itself. That is, we aim to unravel the chemicals that the brain generates to survive. Chemicals of this general type, which make things happen in cells, are termed "signals" or "messengers." Our work has identified novel signals, or messengers, that promote brain cell survival, and we are studying these messengers to see how they can be used for neuroprotection in stroke. This is a step-by-step process, first understanding how the brain defends itself against the consequences of stroke, then figuring out how to exploit those natural, endogenous mechanisms for protecting the brain. We have focused, in particular, on the significance of the essential fatty acid DHA (docosahexaenoic acid), which is present in high quantities in the cell membranes of the brain.

A distinguishing characteristic of the central nervous system is its highly networked organization of synapses and closely interacting cells that results in one of the largest membrane surface areas of all cells of the human body. To understand this, imagine the membrane of a neuron in the hippocampus, spread out on a flat surface. You will see, as part of the large surface area, the extensive branching and complexity of dendrites and dendrite spines (largely made up of postsynaptic membranes). During brain development, many complex proteins and other molecules organize and build these dendrites. The dendrites' shape and length dramatically change during both learning (strengthening) and aging (progressive decline). In stroke, a sudden loss of synapses occurs, while in Alzheimer's disease, the demise of synapses is slow and ongoing. Other brain cells (astrocytes, oligodendrocytes, and microglia) also have very large plasma membrane surface areas. These cells interact with each other and with endothelial cells that line blood vessels in the brain. All of these cells also have extensive intracellular membranes.

Phospholipids (water-insoluble substances, or "lipids," that contain phosphorous) are major structural constituents of these cell membranes, and their role has not been sufficiently appreciated. Small specific pools of membrane phospholipids are reservoirs of potent, biologically active lipids. When cell membranes are stimulated by cell signaling activity, enzymes (called phospholipases) free lipid messengers from these reservoirs. The lipid messengers then regulate and interact with other signaling cascades to contribute to the development, differentiation, function protection, and repair of the nervous system.[4]

This release of lipid messengers has put us on the track of one messenger with great promise in neuroprotection. Some of these messengers are derived from essential fatty acids contained in phospholipids. (Essential fatty acids are those that the human body is unable to make and that, as a consequence, we must obtain from what we eat.) One of these essential fatty acids, DHA (a member of the omega-3 family of essential fatty acids abundant in fish) is more plentiful in the human central nervous system than in any other part of the body. DHA's neurobiological significance is just beginning to be clarified, but we know that it is continuously required for forming and maintaining the structural and functional integrity of membranes of neurons and photoreceptors. DHA is involved in learning, memory, and vision. The brain's supply of DHA is provided by the liver, where DHA is incorporated into the bloodstream's lipoproteins for delivery to the brain cells where it is needed.

Stroke fosters increased oxidative stress in the brain, a process triggered by toxic cell products formed as a result of excessive oxygen. Oxidative stress damages many brain cell components. For example, it impairs the mitochondria, the powerhouse of the cell, and it disturbs the communication between neurons. Oxidative stress triggers chain reactions within the cell and initiates a process known as apoptosis, which, in essence, is cell death from within, because the DNA of genes is chopped off in pieces and the cell falls apart. Since oxidative stress also triggers the release of DHA, our research team turned our attention to whether DHA could tell us how oxidative stress in stroke might be countered.

A Newly Discovered Messenger in the Brain

Using sophisticated technology, we recently detected in the brain the synthesis of a messenger we call neuroprotectin D1 (NPD1)—a messenger made from DHA through oxygenation.[5] (In this instance, oxygen is good, because it is added in a well-regulated reaction, unlike excessive oxygen produced by oxidative stress.) We wanted to find out if NPD1 was a participant in the brain's response to stroke.

To find out, we used a procedure in experimental animals called "middle cerebral artery nylon suture occlusion" (MCA-O) to cut off blood flow in the artery for two hours, producing an infarct—an area of damaged tissue resulting from obstructed circulation—that resembles an ischemic stroke caused by blockage of the middle cerebral artery in humans. This experimental technique also allows a restoration of blood flow resembling reperfusion in human ischemic stroke. Animals that undergo this procedure develop large infarcts and, as a result, have severe neurological deficits.

We discovered that when stroke activates the pathway in the brain described above—by which enzymes free lipid messengers—DHA gives rise to NPD1, and NPD1, in turn, strongly inhibits some of the damage (infiltration of white blood cells and expression of the enzyme COX-2, both of which are part of the inflammatory process) that results from ischemia and reperfusion as well as decreasing the size of the stroke infarct as shown in Figure 1. In the side of the brain affected by the stroke, NPD1 peaks at eight hours, and after twenty-five hours of restored blood flow, it is still elevated, reflecting the ability of the brain to form NPD1 from endogenous DHA. In another experiment, NPD1 was infused using a minipump into the third ventricle of the brain of experimental animals undergoing forty-eight hours of reperfusion after one hour of artery occlusion (MCA-O), and neuroprotection took place. This result demonstrates directly the neuroprotective bioactivity of NPD1.[6]

Further experiments demonstrated that NPD1 acts within the brain cell, before damage to the mitochondria can occur, by serving as a brake

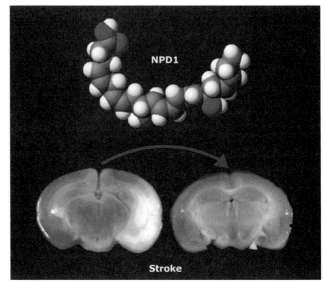

Figure 1   A molecular model of the messenger neuroprotectin D1 that decreases the area of damaged brain tissue (in fact, shown in white in the brain on the left). *Courtesy Nicolas Bazan*

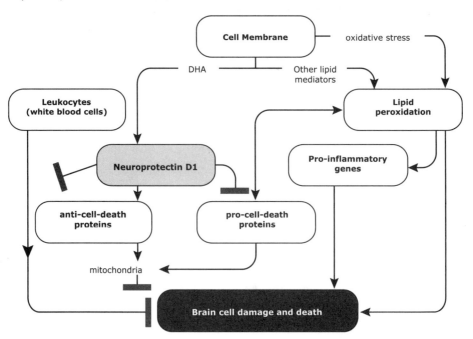

Figure 2   Neuroprotectin D1, formed from DHA, acts as a brake to halt the processes in the brain, caused by oxidative stress such as that resulting from stroke, that lead to cell damage and death. *Courtesy Nicolas Bazan*

for oxidative stress in the cascade of events that leads to cell death. This messenger enhances proteins that promote cell survival and decreases proteins that facilitate cell death, as shown in Figure 2. The cumulative outcome of the actions of NPD1 is protection of cell integrity and function.

Discovery of NPD1 has revealed some of the ways the brain modulates its response to inflammatory injury and provides potential targets for new drugs to treat neurological disorders, in addition to stroke, that have a neuroinflammatory component, such as traumatic brain injury, spinal cord injury, and degenerative diseases such as Parkinson's and Alzheimer's diseases.

Bolstering NPD1 in the Brain After Stroke

Albumin, an important protein in blood plasma (fluid), is able to carry fatty acids in the bloodstream by way of sites on the protein's surface where the fatty acids attach. Our research team decided to take advantage of this transport function to try delivering DHA to the brain after stroke. We reasoned that albumin carrying DHA, if injected intravenously in the peripheral circulatory system, would gain access to the damaged brain, since stroke disrupts the normal blood-brain barrier that prevents many substances from entering the brain. If albumin-mediated delivery of DHA to the brain was successful, it would promote synthesis of NPD1 and hence neuroprotection.

To test this prediction, we used the MCA-O procedure in experimental animals. When we tested the animals seventy-two hours after treatment with DHA-albumin (administered intravenously two hours after the two-hour blockage), the animals' neurological test scores were better than those of animals treated with albumin alone. Infarcts in the cortex of the DHA-treated animals were 86 percent fewer than in animals injected with only a saline solution. Similarly, the size of infarcts in animals that received the DHA-albumin was 70 percent less than in the saline-treated animals. Albumin alone did not significantly reduce infarcts in the part of the brain called the corpus striatum, but DHA-albumin reduced them by 50 percent. In addition, swelling of the brain in animals that received the DHA-albumin was reduced by 58 percent. When we analyzed the animals' brain chemistry 20 hours after the experimentally caused strokes, we discovered a large accumulation of NPD1 in the hemisphere on the same side as the experimental stroke in animals treated with DHA-albumin, showing a correlation between NPD1 and remarkable neuroprotection.[7]

A clinical trial sponsored by the National Institutes of Health is studying the use of albumin alone in human stroke patients. This trial was based on previous research in animals by Myron Ginsberg, M.D., at the University of Miami showing that moderate-to-high-dose human albumin therapy is neuroprotective in models of experimentally induced cerebral ischemia.[8] Initial results of the clinical trial have demonstrated that human serum albumin injected into the bloodstream does elicit neuroprotection. Shortcomings in the procedure were found, however, because high-dose albumin, while effective in protecting the brain, also thickens the blood and may, on occasion, precipitate congestive heart failure. As a consequence, using a lower concentration of albumin is desirable, but this lower concentration is not as effective as a higher one. Thus DHA combined with low-dose albumin, as we have demonstrated, may be a safe and effective approach to protecting the brain after stroke.

Hope for Neuroprotection

The brain responds to any kind of injury with a plethora of signals, some of which are harmful and some of which are protective; the preponderance of signals in either direction influences the overall outcome (neuronal survival or death). Such is the case with inflammatory and counter-inflammatory lipid messengers. Therapeutic strategies targeting neuroprotection will mount the first truly effective defense against brain damage in stroke and very possibly prove to be the most effective approach to other central nervous system disorders with a neuroinflammatory component.

Stroke depletes discrete specific pools of DHA from the brain, first, because ischemia releases free DHA within the tissue and, second, because the surge of oxygen when blood flow is restored activates excessive oxidative stress and consumes more DHA. But lipids such as DHA are involved in maintaining the structural and functional integrity of brain cell membranes, and therefore replenishment of DHA after a stroke may be required. DHA from what we eat is essential for this, but most diets do not contain adequate amounts of fish, which is rich in omega-3 fatty acids, including DHA. For reasons we do not yet fully understand, DHA cannot be forced into the brain during a short period of, for example, a diet based predominantly on fish. Therefore, a balanced diet with plenty of fish should be the norm throughout life.

As a therapy in acute stroke, delivery of DHA to the brain—for example by injecting albumin that carries DHA—will provide the brain with the precursor for NPD1 and enable it to rebuild critical phospholipid pools depleted during ischemia and reperfusion. Still to be defined are what signals turn on the formation of NPD1, and we must also learn more details about NPD1's mechanism of action in rescuing brain cells from oxidative stress. This information, in turn, may provide a template for developing new neuroprotective drugs. The brain, after all, knows how to defend itself, and drugs emulating this natural ability could be powerful weapons in the battle against the destruction caused by stroke.

Although research on neuroprotection mechanisms and their application, whether for stroke or for other devastating diseases such as Alzheimer's, has in many ways been disappointing so far, we are learning more each day about the cellular and molecular mechanisms involved in these diseases. At the same time, a new emphasis in translational research is bringing neuroscience discoveries made in the laboratory closer to the patient, such as Tasha with whose story we began.

CHAPTER 2

Why Not a National Institute on Pain Research?

Kathleen M. Foley, M.D., *with* Maia Szalavitz

Kathleen M. Foley, M.D., is an attending neurologist in the Pain and Palliative Care Service at Memorial Sloan-Kettering Cancer Center, as well as professor of neurology, neuroscience, and clinical pharmacology at Weill Medical College of Cornell University. Dr. Foley holds the Chair of the Society of Memorial Sloan-Kettering Cancer Center in Pain Research and is medical director of the National Public Health Palliative Care Initiative of the Open Society Institute. She is a past president of the American Pain Society. Dr. Foley's work focuses on the treatment of patients with cancer pain and advancing palliative care. She can be reached at foleyk@mskcc.org.

Maia Szalavitz is a freelance writer on neuroscience and addictions whose articles have been published in the *New York Times*, the *Washington Post*, *New York Magazine*, and *Newsweek*. Her most recent book is *Tough Love America: How "Boot Camps" and Other "Get Tough" Programs Fail Teens and Families* (Riverhead Books, 2005). She can be reached at maiasz@gmail.com.

WHEN I BEGAN my clinical training in neuro-oncology at Memorial Sloan-Kettering Cancer Center in the 1970s, little was understood about how the brain processes and feels pain. For example, fewer than ten neurotransmitters of any kind were known to exist in the entire brain, and only a small portion of their receptors had been identified. No one knew how opioid medications like morphine worked. The brain's own pain-killers and the receptor systems on which they acted were just starting to be discovered.

Attitudes toward patients in pain—even those who were dying—were just as primitive. Patients and physicians alike thought that pain should be viewed as a sign that could help the doctor find out what was wrong, not as a symptom or a disease entity that should be targeted in itself for relief. Addiction to pain medication, believed to be almost inevitable if such medication was provided, was considered far worse than some "discomfort."

As a young clinician, I was asked to help start the first pain service in an American cancer center, which formally opened at Memorial Sloan-Kettering in 1981. Our clinical work and the work of our laboratory colleagues has produced much new knowledge since then, but, unfortunately, myths about pain and its treatment still thrive and are being reinforced by recent news reports about prescription drug misuse and prosecutions of doctors for "overprescribing." We need far more research in this area to produce better medications, but the almost fifty million Americans suffering chronic pain need better care now.

Here, I will share some of what I have seen and learned about treating pain and highlight where more research—and greater understanding by the public and regulators—is urgently needed.

In the Beginning

I did not make the connection with my life's work until much later, but my first experience of cancer pain was with my mother, who died when I was in my early teens. She never had good pain management, and I certainly observed that, but I completely disconnected it from what eventually became my field of research.

In my early medical training, I had picked up the negative attitudes of other doctors toward treating pain, but at Memorial Sloan-Kettering I was seeing the burden of pain in cancer patients and the great potential that lay in understanding its cause and alleviating it. The compelling aspects of cancer pain drew me to the search for better therapies. Our clinical studies focused on doing placebo-controlled, randomized studies of old and new opioid drugs and drug combinations to establish safe and efficacious doses and define guidelines for their use. One of our researchers, Gavril Pasternak, M.D., Ph.D., now head of molecular neuropharmacology at Memorial Sloan-Kettering, had been involved in the critical early work of identifying and elucidating the action of opioid receptors in the brain while an advanced doctoral student at Johns Hopkins.

I had learned all the myths about opioid medications, but without recognizing them at the time as myths. I thought that they were pharmacologic facts. I thought that the drugs inevitably caused addiction if they were used long term, that high doses increased the risk, that tolerance would continue to escalate, and that over time the drugs would not even be effective for the pain. Then, I began to see what actually happened when we treated cancer patients in pain with opioids. I saw how the medications dramatically improved their lives. Almost none of the patients developed addiction or what is referred to as "psychological dependence." Higher doses were needed for more severe pain, higher sensitivity, or larger people. Tolerance, when it did develop in some cases, came about as a result of the patient's progressive disease with escalating pain. Increasing the dose was effective in relieving the pain. In short, we could not demonstrate a clear tolerance to analgesia in this patient population. Importantly, however, tolerance—that is, diminishing effect—usually developed rapidly to the sedation and mild confusion, as well as to the potentially deadly respiratory depression, that opioid medications can cause.

Addiction: Battling the Myths

Our extensive experience treating cancer-pain patients with chronic opioids pointed toward what we later learned about the multitude of endogenous (natural) opioid receptors: different processes were involved in the development of tolerance to different effects. Researchers have also found that the development of physiological need for the drug (dependence) and consequent withdrawal symptoms involved different processes than did the development of psychological drug craving and compulsion to take the drug despite negative consequences (addiction). These differences seem increasingly likely to involve receptor or neurochemical differences in the brain. Almost all of our patients developed physical dependence if they took the drug for weeks or longer, but, when the cause of the pain was effectively treated, they did not want to continue taking the drugs.

I have seen thousands of patients managed well with long-term use of opioids. The only patients who developed drug-use problems were those who had a prior history of drug addiction. Because many of these patients were seriously ill with cancer, we worked out special programs to address both the pain and the addiction behaviors of this group.

I remember one patient, a woman in her forties, who had severe pain from a large lung tumor invading her upper chest and the nerves to her right arm. She was taking high doses of opioids without effective relief. She had a surgical procedure known as a cordotomy, which, by means of a small lesion made in the high cervical spinal cord, interrupts the pain pathways in the spinal cord that go to the brain. Following the procedure, she had no pain and refused to take her pain medications, because she said that she did not need them. Because she refused to taper off the drugs, she suffered withdrawal; but the flu-like symptoms such as abdominal cramping and diarrhea did not trouble her significantly. She simply had no psychological dependence on the medication.

One of the hardest groups to persuade to allow treatment with opioids were the parents of children with cancer pain. Some parents feared the medical meaning of these drugs—thinking that if their child took morphine, it meant that the child was dying—and could not accept that possibility. Others were so frightened of addiction that they refused permission to give the medications, even when their children were obviously in pain. As our clinical experience and the experimental evidence for the safety of these drugs in cancer treatment grew, we recognized that we needed to educate both doctors and patients about it.

One fourteen-year-old girl and her family let us make a video of her story. She had been treated for a bone cancer in her leg, and the surgery required to realign her bones had left her with severe pain. Her surgeon had told her family that she should tough it out, because the operation should not have caused such a high level of pain. When we saw her, she was in agony, crying constantly, curled up in a fetal position and completely unable to function. If she was ever going to be able to walk again, she would have to exercise. We needed to get her pain under control for her to begin rehabilitation.

We gave her methadone and witnessed a transformation that was remarkable. She became a smiling, appropriate fourteen-year-old girl again. Her parents were amazed. During about five months of rehabilitation, she remained on tapering doses of methadone and was able to do her exercises and go back to school. Following the completion of her rehabilitation, she was walking again and the drug was stopped. She did not have any problems with either taking it or stopping it.

During the past three decades, we have had thousands of such experiences. We even studied heroin in patients, because heroin was and still is used in the United Kingdom for pain treatment and some groups in the United States were advocating it as better than morphine. We found that, for most people, a wide range of opioids worked equally well and so decided that the political battle to legalize heroin for pain treatment in the United States was not the issue for us. Rather, we needed to address the barriers to adequately treating pain with the available opioids.

We also advocated for patient-controlled use of pain relievers. When we first started this, the technology did not exist that is now used to allow patients to push a button to infuse the drugs. We decided to let patients keep pills at their bedside and take them as needed, avoiding the timing problems associated with the vagaries of nursing staff availability. Some people thought that this was a sure recipe for addiction or overdose, believing that patients would simply take as many drugs as possible, as quickly as possible. In fact, however, cancer patients tended to take smaller doses, and often fewer doses, when they knew that they could control their own pain. Since then, psychological research has shown that a greater sense of control over pain actually diminishes the sensations of pain, and patients are effective in timing the administration of their medications, balancing pain relief and side effects.

Unfortunately, news about our innovative practices prompted a visit from a New York State narcotics agent, who advised us that allowing patients to administer their own doses in this way was illegal. We stopped, of course, but this illustrated how law enforcement and regulation can short-circuit promising new approaches in drug administration. Today, patient-controlled delivery of opioid drugs using infusion devices has become standard practice, and better control of pain consistently has been linked to faster healing and better recovery.

The Experience of Cancer Patients

Back then, almost every notion promulgated about opioid drugs was wrong, and indeed almost completely backward: the notion of a high likelihood of addiction, the idea that pain was important to healing and that blocking it might make matters worse, the idea that the opioids invariably lose effectiveness. Over the years, clinical experience and data have shown that in the cancer population there is a very minimal risk of addiction, even with large exposure to opioids, and the benefits to patients are undeniable.

The cancer-patient experience in the United States, together with the influence of a worldwide hospice movement, provided the clinical framework for providing pain relief as an essential aspect of humanizing end-of-life care and helping patients die with dignity. Better pain relief reduced patient requests for assisted suicide and eliminated needless suffering.

National and international surveys report that more than four-fifths of cancer patients have significant pain in the last months of life, pain that requires treatment with strong opioids. The World Health Organization (WHO) Cancer Unit took up this issue in its campaign "Freedom from Cancer Pain," calling attention to the need for opioids to be available as essential drugs. Embedding the need for pain relief into its international cancer-control strategies, WHO emphasized the need to use the knowledge that we have today to improve the quality of life for dying patients in pain.

Despite WHO's publication of four monographs describing the need for pain and symptom management and their integration into health-care programs for cancer, despite the development of a list of essential drugs for pain and palliative care and a host of directives to governments, the majority of people in medium- and low-resource countries around the world have no access to opioid drugs for essential pain relief. According to WHO, developing countries account for four-fifths of the world's population, but access 5 percent of the world's opioids. In about half of all countries, opioid medications are rarely used, even for the dying. This has stirred new interest in the need to provide pain and symptom management to patients dying with HIV/AIDS, more than half of whom in the last months of life have significant pain requiring opioid pain relievers.

Unfortunately, concerns such as addiction, coupled with lack of professional education and a low priority for the care of seriously ill or dying patients in pain, continue to limit the availability of opioids around the world. Although there are model programs in India providing morphine to cancer patients and in Uganda for both cancer patients and HIV/AIDS patients, there is still a great need to advance and encourage balanced drug policies that reduce the regulatory barriers and restrictions that limit suffering patients' access to cheap, cost-effective pharmacologic pain treatment.

As the role of opioids in treating cancer pain evolved, attention turned to the potential of long-term or chronic opioid therapy for patients with chronic non-malignant pain. Clinical studies pointed to the efficacy of opioids in a spectrum of pain states from arthritis to pain that persists after a case of shingles, what is known as post-herpetic neuralgia. Nerve injury pain (neuropathic pain) had traditionally been viewed as resistant to opioid analgesia, but, again, carefully controlled studies demonstrated responsiveness to opioids, albeit often at higher doses than required for somatic (body) pain states.

At the same time, activists and physicians in the United States were pressing for better recognition of the need for pain control. The Joint Commission on Accreditation of Hospitals, designating pain a "fifth vital sign," now requires health-care institutions to assess pain levels of patients, manage pain, and provide patients with information about available pain therapies. Physicians, patient advocates, and pharmaceutical firms began to argue for the need to treat chronic pain and question why opioid drugs would be denied those whose suffering could last decades. Moreover, research has shown that widening the use of prescription drugs such as morphine and fentanyl, mostly for cancer pain, was not associated with an increase in misuse.

Misuse and Misperception

Early in this decade, however, Oxycontin (a time-release formulation of oxycodone) began to be marketed for patients with moderate-to-severe chronic pain, and, between 2000 and 2002, there was a more than fourfold increase in abuse cases reported to hospital emergency rooms. As media reports described how abusers use the drug to get high, addicts readily learned how to defeat the time-release mechanism by snorting, injecting, or swallowing the drug. The high-dose formulation made these methods of taking the drug extremely intoxicating and potentially deadly to naïve users such as experimenting teenagers. What actually gave rise to the epidemic of abuse is still being studied, but the causes likely include diversion of the drug from appropriate channels by means of drugstore robberies, illegal prescribing by some physicians, and diversion by patients. The problems seemed to occur most commonly in rural areas, such as Kentucky, West Virginia, and Maine.

It is still unclear to what extent illegal prescription abuse has anything to do with appropriate prescribing for patients with pain, but concern about the growing incidence of prescription drug abuse has led to a greater focus on physician prescribing. A series of high-profile cases of physicians accused of inappropriately prescribing opioids, and, in some cases, causing patient deaths, has added to the concern about prescription fraud and misuse. Law enforcement bodies such as the Drug Enforcement Agency (DEA) are holding doctors criminally responsible if their patients fake complaints, misuse drugs, sell the drugs, or overdose on them. These prosecutions are having a chilling effect on clinical practice, as patients who need large doses bring government attention even when they have a documented pain condition such as multiple sclerosis, failed back surgery, or spinal cord tumors. Testimony in these cases has demonstrated a wide range of opinions on standard doses, the role of opioids, and standards of treatment. What is of great concern is that courts are trying to define what should be standard medical practice—not only as a matter of legal judgment but as a matter of medical oversight.

Along with my colleagues, I participated in a DEA project to develop a "FAQ" (Frequently Asked Questions) to educate doctors about appropriate prescribing of opioids. The FAQ was published on the DEA Web site in August 2004, but was then withdrawn by the DEA, which claimed it contained inaccuracies—despite extensive collaboration and vetting of the document with leading medical experts on pain. Withdrawal of the FAQ also coincided with its use in the defense of physicians being prosecuted for their opioid-prescribing practices. A group of six past presidents of the American Pain Society, including me, sent a letter to the judge in one such case, disagreeing with expert testimony obtained by the prosecutor about the use of opioids for chronic pain. This expert testimony apparently was considered crucial in deciding some aspects of the case, but it was based on neither evidence nor standard practice.

When a physician is arrested or loses his license to prescribe, his patients are abandoned, their records confiscated, and no provision made for the ongoing care of those patients taking opioids. This lack of concern for the patients is at best unprofessional and clearly unethical, yet it has recently affected hundreds of patients in Pennsylvania, Virginia, and Montana.

In the Brain: Very Real But Very Different Responses

Ironically, this backlash against treatment with opioids comes at a time when we are understanding both pain and addiction better than ever before. Neuroscientists have now identified more opioid neurotransmitters in the brain than the total of all transmitters of all types known before 1970. For example, research has identified three different families of opioid receptors with multiple subtypes; the first such transmitter was isolated only in 1973. In just one class, the mu-receptors, more than two dozen different subtypes have been cloned in the mouse, ten in humans, and, according to Dr. Gavril Pasternak, "there are almost certainly more."

Pain is processed in a network of brain systems called the "pain matrix." This matrix comprises cortical areas near the surface of the brain, emotional regions toward the center of the brain, and sensory areas in several locations. But physical pain and emotional pain are not at all distinct from each other in the brain. Psychological factors like how helpless a patient feels are associated with greater pain processing in sensory regions, not just areas linked typically with emotion. On functional magnetic resonance imaging (fMRI) scans, both the emotional pain of losing a relationship and sensations of physical pain light up an area called the anterior cingulate cortex.

Meanwhile, addiction researchers are discovering that people who develop drug-abuse behaviors tend not to be pleasure seekers but people who have suffered trauma, such as child abuse, or have had a mental illness, such as depression. Scientists are also finding genetic factors that appear to predispose people to drug problems.

Dr. Pasternak calls attention to the remarkable diversity of human responses to opioid medications. These responses are so variable that about half of all clinical trials do not find morphine to be an analgesic. He explains: "This is not because morphine is not an analgesic—it is—but because there is so much 'noise' due to genetic variation that you have a hard time showing it." He adds that, in rodents, when the researcher uses just one physiologically consistent strain, the data are much cleaner and more obvious.

Opioids are unique in several ways that at least until now have made them indispensable to patients. Unlike acetaminophen, aspirin, or non-steroidal anti-inflammatory drugs (NSAIDs) such as ibuprofen, opioids work against pain by means of at least two different mechanisms. In the spinal cord, opioids appear to work directly by preventing some pain messages from reaching the brain at all. In the brain, however, they act even more strongly to combat the distress associated with pain. Patients frequently describe the drugs as distracting them from their pain, saying things like "I still feel the pain, but it doesn't bother me now."

Although the drugs do decrease pain intensity by their spinal cord action, the relief they bring by literally "de-stressing" a patient appears to be greater. No other pain medications have this action and there is no pharmacologic ceiling on the pain relief that they can provide. All of the other commonly used painkillers are limited by their mechanisms of action to treating mild to moderate pain. Even if they did not cause potentially dangerous side effects, such as bleeding in the stomach at higher doses, this limit on their range of effectiveness curtails their usefulness. By contrast, there is no upper limit on opioid dose, providing the patient can tolerate side effects such as drowsiness and the drug's effect on breathing. If you increase the dose, you increase the pain relief.

"To date, opioids have proven to be the most efficient and effective of all of the different courses of drugs," says Dr. Pasternak. He notes that there are alternative painkillers that look great in the laboratory, with potency equal to or greater than morphine, but when these drugs reach clinical trials, they all have intolerable psychiatric side effects. As he puts it, "they make you crazy." Many such drugs, however, act on the brain's NMDA receptors. Animal research had suggested that by blocking these receptors, tolerance to opioids could be reduced or even eliminated, along with withdrawal symptoms. One weak NMDA-receptor blocker, dextromethorphan (sold as an over-the-counter cough suppressant) seemed promising for use in conjunction with opioids. Unfortunately, it turned out that doses of dextromethorphan high enough to affect tolerance and dependence were also high enough to produce hallucinations and other unmanageable side effects. Thus this particular approach did not pan out.

Calling for a National Institute on Pain

Despite their effectiveness, the opioids remain but part of the pharmacologic approach to managing patients with pain. Researchers are focused on teasing out the molecular aspects of the various kinds of pain mechanisms (somatic, neuropathic, and visceral) and seeking better understanding of the potential of other receptor systems for modulating pain relief. For example, there are exciting studies of the role of sodium, potassium, and calcium channels; the potential role of cannabinoid receptors in analgesia; and the role of targeted toxins.

In recognition of the volume and importance of this research, the first decade of the twenty-first century was designated the Decade of Pain Research. And yet, notwithstanding the obvious potential for new discoveries and the urgent need for better therapies, less than one percent of National Institutes of Health (NIH) grants are made for primary pain research. That is despite the fact that about one in five Americans suffers serious chronic pain, a number that rises to almost half of Americans in older age groups.

In contrast, about 15 percent of Americans sometime in their lives will have a substance-abuse problem at some level from mild to severe, but five times more money is spent for research on drug abuse than on pain. In fact, two entire institutes, the National Institute of Drug Abuse (NIDA) and the National Institute on Alcoholism and Alcohol Abuse (NIAA), are devoted exclusively to research on substance abuse and misuse. There have been many efforts to begin to achieve a better balance: repeated NIH workshop recommendations, several Institute of Medicine recommendations, and two State of the Science Meeting recommendations, for example. But we still have no organized government support for either centers of excellence in pain research, which could support clinicians and basic researchers in advancing pain therapies, or an institute to create the necessary leadership to advance pain research and policy in America and to address what has been recognized as a serious public health issue.

What is needed is a National Institute on Pain, with a budget at least equivalent to the budgets of the institutes on addictions. For, like addiction, pain affects not only the person with the disorder but also his family and his work life. Like addiction, pain robs life of joy and meaning. The outlook for productive research on pain and pain management has never been brighter. A National Institute on Pain could draw the public's attention to these points, while informing and educating people about opioid medications. It could convey the message, for example, that addiction occurs when a drug makes a person's life worse, but for pain patients, the very same drug can make life infinitely better. Indeed, a National Institute on Pain could bolster support for addiction treatment as well, by making the research-backed case, along with NIAA and NIDA, that addiction is a treatable disorder, serious but not worse than a death sentence—and not worse than chronic agony. This research could build on long clinical experience with cancer patients to demonstrate conclusively, by means of long-term studies of people with both cancer and chronic non-malignant pain, that addiction is rare in this population.

We also need better medications, because some pain opioids can relieve only partially and some patients cannot tolerate the doses that they need to get relief. We need medications and other therapies that can help those for whom opioids fail entirely. And we need to fine-tune the opioids so that we can match particular drugs with patients for whom they are best suited and avoid giving them to patients for whom they would be ineffective—or likely to have serious side effects because of a patient's genetic predisposition. For example, about one out of ten Caucasians cannot metabolize codeine, making it for them an ineffective analgesic. Also, because tolerance to various effects such as physical dependence and addiction are separate processes, it may someday be possible to create medications that avoid or reduce problematic processes while maximizing those involved in pain relief. More research on dissociating these effects is needed.

In the past thirty years, we have seen at least some barriers to pain relief—for example, for dying patients—fall dramatically. We are far more aware of the impact of pain on patients and families and of the enormous need for providing approaches that focus on the physical, psychological, social, and spiritual needs of patients and families. Clearly, research on opioid analgesics is not going to be the only solution for all patients with pain. Yet, the role of the opioids today is so central that we need more research to better understand who can benefit from these agents, what risks they impose, and what other drug combinations are effective. Until we have achieved that, though, does it seem reasonable that patients are forced to endure needless suffering because legal or regulatory barriers make existing medications unavailable to them?

CHAPTER 3

A Brain Built for Fair Play

Donald W. Pfaff, Ph.D.

Donald W. Pfaff, Ph.D., is a professor and the head of the Laboratory of Neurobiology and Behavior at the Rockefeller University, where he and his research group study how the mammalian brain manages specific natural behaviors and the hormonal and genetic influences on brain arousal. His most recent book is *Brain Arousal and Information Theory* (Harvard University Press, 2006). He can be reached at pfaff@mail.rockefeller.edu.

ASSOCIATED WITH EVERY RELIGIOUS SYSTEM I have read about, across continents and centuries, is a rubric, known in Christianity as the Golden Rule, that requires me to do unto you as I would have you do unto me. This rule is so ingrained in how we behave toward each other that we rarely stop to question its source. If pressed, we might speculate that its origins are lost in the mists of time—for example, when the first high priests figured out how to satisfy a sovereign's demand for social stability. But suppose, for the moment, that the Golden Rule is even older, that it is as old as our own biology. Suppose that although the Golden Rule might have acquired all sorts of sociopolitical decorations over the course of human history, it nonetheless is traceable to identifiable processes in our brains. If this were so, we could understand why this rule and its many variations have survived so widely in human ethical systems, philosophies, and religions.

In his recent book, *The Ethical Brain*, Michael S. Gazzaniga, Ph.D., challenged us not only to face the ethical questions inherent in neurobiology but also to try to understand how our brains govern our ethical responses themselves. Here I want to explore a new theory of the neuroscientific basis for the human instinct for fair play.

Fair Play: Human, Animal, and Computer

The evidence for this universal principle of fair behavior is almost overwhelming; it comes not only from religion but also from anthropology, archaeology, the study of animal behavior, and even computer science.

In their book *Inside the Neolithic Mind: Consciousness, Cosmos, and the Realm of the Gods* (Thames & Hudson, 2005), South African scholars David Lewis-Williams, Ph.D., and David Pearce, Ph.D., argue from archaeologic and anthropologic evidence that neural patterns of activity hardwired into the human brain help explain the range of religious art and social practices they believe were produced and exhibited by Neolithic people. They write, "The commonalities we highlight cannot be explained in any other way than by the functioning of the universal human nervous system." Similar images in the art of widely separated tribes suggested overwhelmingly to those authors that a common mentality underlay their religious beliefs. Biologic forces in the form of common neuronal patterns of activity have helped shape human behavior in society, including learning not to treat others in ways we would not want to be treated ourselves.

My suspicion that basic neurobiological mechanisms foster fair play is deepened by realizing that animals also behave this way. Animal behaviorists call it "reciprocal altruism." Individual animals take risks to protect the group; one animal does things that help other members of its group survive, just as it would want those other members to act for its benefit. A small bird sounds its alarm call in a manner that alerts and saves the rest of the flock from the approaching hawk, even though by doing so it reveals its own location to the hawk. A baboon shares food with a hungry member of its troop, even though by doing so it has less food for itself.

Moreover, the mechanisms involved are not mystical; even computers can be programmed to behave "fairly." Robert Axelrod, Ph.D., and William Hamilton, Ph.D., at the University of Michigan, programmed computers to display mutual cooperation and, by running a computer tournament, showed that such behavior is an evolutionarily stable strategy. In other words, mutual cooperation in these computer games started spontaneously, thrived, and resisted opposition. They started the computer tournament with each "player" able to exhibit either mutual cooperation or selfish defection by following these two rules: 1. "On the first move, cooperate" and 2. "On all subsequent moves, do whatever the other computer did on the first move." Thus, all computers cooperated on the first move and, if they followed the second rule, also cooperated on subsequent moves, a "tit for tat" strategy.

If computers can display cooperative responses, then we know that straightforward physical mechanisms must be sufficient to explain the result. In the research by Axelrod and Hamilton, an ongoing interaction between computers was necessary for cooperation to thrive. Human beings, too, are more likely to treat another person properly if they expect to interact with that person again. We obey the Golden Rule when we sense that we inhabit the same "space" as the other person. Conversely, we do not murder, because we do not want to be murdered ourselves. We see that the people to whom we would do harm actually share fates with us.

Fear, from the Thalamus to the Amygdala

Evidence from every direction points to an underlying neurobiology equally stable over time and across cultures and species—yes, even including the computer. What I propose is that the mechanisms in our brains that guide behavior adhering to the Golden Rule are those that govern our fear responses. I further propose that the process involves a blurring of identity, in which one person's envisioned fates and fears are merged with another's. We share not only our fates but our fears with others. To understand why this might be so, I should explain a little about the complicated processes involved in fear, from the neural pathways in the brain to genes and hormones, emotions and memories.

Some stimuli from the environment produce fear, others do not. The difference in our brains derives from the ability of fearsome stimuli to trigger activity in an ancient part of our forebrain called the amygdala. There, certain genes and a few messenger chemicals are the crucial actors. A series of landmark discoveries in several laboratories has shown how stimuli produce fear.

The two basic kinds of fear, unlearned and learned, depend on the amygdala—at least ten ancient groups of neurons lying roughly parallel to our ears—which participates in primitive neural circuits deep in the forebrain. Unlearned fear, often manifest by animals stopping all movements ("freezing") to see what is going on and to assess their risks, depends on a cell group called the central nucleus of the amygdala. Joseph LeDoux, Ph.D., and his colleagues at New York University showed that learned fear depends on storage of an emotional memory trace in a different part of the amygdala, the lateral amygdala. To paraphrase James McGaugh, Ph.D., at the University of California, Irvine, neurons in the lateral amygdala make emotionally significant events memorable.

Scientists argue about what fears are innate, that is, unlearned. Some think that human beings are born with the fear of only a small number of events, such as loud noises and falling. Learned fear, however, develops when emotionally neutral stimuli and neutral environments—those with no particular emotional meaning—are associated with innately fearful experiences such as severe pain or other terrible events. For example, we could learn to fear a particular kind of house if we were subjected to deeply frightening experiences in a similar house.

Stimuli that produce fear are not recognized as fearful from the microsecond they touch our sensory receptors. Instead, they enter our nervous systems value free, through the skin and up the spinal cord, on our tongues or through our noses, from our eyes or ears, and toward the forebrain. They reach a forebrain region called the thalamus. At that point, signals split and travel to the amygdala by two different routes, as shown in Figure 1.

One set of signals ascends toward the sensory areas of the cerebral cortex. In fact, the word "thalamus" comes from the Greek word meaning "antechamber" because, neuroanatomically, the thalamus acts as an antechamber to the cortex, that magnificent covering of our brains. From the cerebral cortex, those signals travel to the amygdala. The other set of signals goes directly to the amygdala from the thalamus. LeDoux's laboratory has shown that the messages that travel these two routes are not identical to each other, and currently neuroscientists endeavor to explain how and why. I would like to posit an answer.

Let me divide all responses that we call "fear" into primitive emotional reactions and more sophisticated cognitive reactions. I theorize that innately fearful events trigger electrical discharges in nerve cells located in the amygdala, straightaway, having used the direct route from the thalamus. These electrical signals lead to our experience of emotional fear. Learned fear, involving memories of initially neutral events that have become associated with fear, is more complex. I propose that our

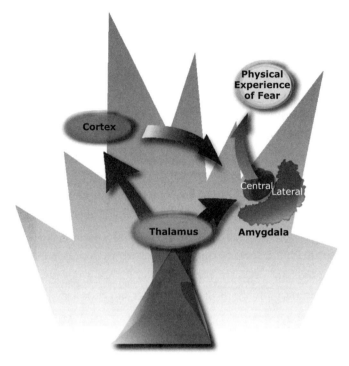

Figure 1 After signals related to fear reach the thalamus, they take two routes in the brain. One set of signals goes directly to the amygdala, while other signals travel first to the cerebral cortex for cognitive processing and then to the amygdala. The first route may relate to our primitive emotional reactions to fear and the second to our fearful thoughts. *Credit: Courtesy Susan Strider, The Rockefeller University*

fearful thoughts depend on the information-processing power of the cerebral cortex.

To appreciate the emotional component of fear, let us first concentrate on the amygdala itself. Michael Fanselow, Ph.D., at UCLA, and others have shown that its lateral parts encode memories of stimuli associated with aversive events. Then, from the amygdala, emotionally loaded signals are sent to many other parts of the forebrain, in particular an ancient part of the forebrain called the septum, and to the hypothalamus. These forebrain cell groups are both necessary and sufficient for the unique experience of fear. A more modern part of the forebrain, the frontal cortex, also receives fear signals, but it functions primarily to suppress sheer fright. For a person to avoid panic and long-term conditions such as post-traumatic stress disorder, the frontal cortex neurons must function normally. In addition, outputs from the amygdala also influence memory processes in many brain regions.

Learned fear employs the same pathways, except that the stimuli associated with negative experiences start as ineffective for activating the amygdala. Only after repeated associations with pain and other negative events do these stimuli produce a reaction in the amygdala.

These circuits in the central nervous system, whether for unlearned or learned fear, produce responses that allow an animal or a human being to avoid the source of the fear. Some of these responses, such as those I have just described, are very rapid. But in addition to these quick electrical and behavioral responses, brain mechanisms for fear make use of slow effects that involve gene expression. These slow effects depend on genes that make large proteins that bind stress hormones and on genes that code for a small molecule responsive to stress, corticotrophin-releasing hormone (CRH). As an intriguing aside, both of these modes of action—rapid/electrical and slow/genomic—provide possible targets for pharmaceuticals that could be used therapeutically to reduce fear.

Three Steps in an Amygdala Neuron

What happens in amygdaloid neurons that constitutes the mechanism of fear? In neurochemical and genetic terms, at least three steps can be identified: the arrival of specific chemical signals at the amygdaloid neuron; the signaling within the neuron consequent to that arrival; and the eventual effect of that signal, whether it is an electrical discharge or a change in gene expression.

Let me give one example. The neurotransmitter called glutamate is involved in fear signaling. As soon as glutamate binds to its receptors on the cell membranes of amygdaloid neurons, calcium is permitted to enter the neuron. Then, the calcium unleashes a long series, what scientists call a "cascade," of complex signals within the neuron. This cascade uses specialized proteins that "decorate" neurons with appendages containing phosphorus and oxygen. Such a cascade is required for the memory of fear.

A second example uses a different type of biochemical cascade. A neuropeptide called "brain-derived neurotrophic factor," known to participate in stress and fear, binds to its specific receptor on amygdaloid neurons. Once binding occurs, a small set of proteins—called "G-proteins" because they are organized around the chemical guanine—change their state of activation. That change triggers a new set of biochemical reactions that, as in our previous example, add appendages containing phosphorus and oxygen to the neurons.

So far I have covered two steps: the arrival of particular neurotransmitters and neuropeptides at an amygdaloid neuron and the biochemical signaling within the cytoplasm of that neuron, beneath the membrane but outside the cell nucleus. The third step determines the consequences of this signaling, which can be either fast or slow. The fast effects are triggered if the last reaction of the biochemical cascade phosphorylates (adds phosphorus and oxygen to) a protein that participates in an ion channel in the amygdaloid nerve cell membrane. That process causes a rapid electrical change in the cell. The slow actions occur in the cell nucleus, where the last reaction of the biochemical cascade would phosphorylate either a genomic transcription factor or a chromatin protein that is part of the covering of the cell's DNA. Therefore, both neurotransmitters and neuropeptides make use of series of phosphorylation reactions to transmit their signals both to membrane ion channels (for rapid electrophysiological changes signaling fear) and to the cell nucleus, for slow adaptations in the genes related to enduring fear states.

Hormone and Gene Connections

The neurotransmitters and neuropeptides that I have discussed affect the physiology of the amygdala. But to complete the story, I must add a fascinating—and complex—hormonal connection. Consider corticotrophin releasing hormone (CRH), mentioned earlier. CRH, which is produced by neurons in the central nucleus of the amygdala, not only fosters behavioral responses to fear but also triggers the stress hormone axis that runs from the hypothalamus to the pituitary to the adrenal gland.

When this axis is triggered, steroidal hormones such as cortisol pour out of the adrenal gland and help the entire body deal with stress and fear.

But the effects of cortisol on behavior depend on whether levels of adrenal stress hormones are chronically and repeatedly elevated. If they are not, the effect of cortisol is restorative, bringing the body's systems back from an emergency state toward a normal one. For example, such stress hormones acting in the rat amygdala would be essential for the extinction of a learned fear response. But if an animal was subjected to chronic fear and stress, such that the adrenal hormones were called on again and again, then cortisol could have the opposite effect, actually amplifying behavioral responses to fearful stimuli. In humans, hormonal mechanisms that originate with CRH in the amygdala can either relieve our stress or heighten our anxiety, depending on our emotional history. The implications of these complex dynamics for the management of anxiety states, post-traumatic stress disorder, obsessive-compulsive disorders, and other illness are clear.

New research continues to discover genes that affect how fear operates in the amygdala, but scientists do not yet know exactly what these genes do in amygdaloid neurons. For example, in a recent paper in *Cell*, a team led by Gleb Shumyatsky, Ph.D., at Rutgers University, reported a gene coding for a protein named stathmin that helps to make a mouse timid. Removing that gene through a biochemical process in the laboratory made the mouse more daring. Such mice failed to avoid innately fearful environments and had less memory of fearful events. In cellular terms, how this happens is still unknown, but electrical recordings from cells in the lateral amygdala showed deficits in responses to inputs from the cortex and the thalamus.

For frightening emotions and emotional memories to operate correctly, in a biologically adaptive fashion, the entire central nervous system must be aroused. In addition to this "generalized arousal," James McGaugh, Ph.D., and his colleagues at the University of California, Irvine, have reported that the proper operation of amygdaloid mechanisms related to fear depend on two specific arousal-related neurotransmitters, norepinephrine and dopamine. These chemical messengers coming from the lower brain stem energize amygdaloid neurons so that they can properly process inputs from the thalamus and cortex related to fear. An average person, or a "jumpy" one, would have enough norepinephrine and dopamine coming from the lower brain stem to the amygdala that he would respond briskly to fearful stimuli. But a person without enough norepinephrine and dopamine might not be afraid at all, even though signals from the thalamus and cortex tell his amygdala that he should be. This scenario may correspond to our experience of the proverbial unflappable "cool guy" who does not seem to react dramatically to situations that deserve a full-blown fear response.

How Shared Fears Make Fair Play

Based on the neurobiology of fear I have explained, I theorize that fear, learned and unlearned, helps a person avoid acts of violence and other behavior that could harm someone. That is, the genes and neural circuits that manage fear provide the crucial biological components of a process that leads to behavior obeying this universal ethical principle. This process is, of course, not all of ethics, but it is certainly central to the explanation of how human beings behave according to one important principle.

From a scientific point of view, the best explanations of complex phenomena are the most parsimonious. Scientists try to use as few and as simple explanatory forces as possible. By referring to the basic, almost primitive, brain mechanisms governing fear, I can achieve this type of scientific explanation for reciprocal altruism. The explanation does not require fancy tricks of learning and memory but instead invokes the easiest step of all: the loss of information—a term I have coined for the blurring of identity between the actor and his target, a distinction that his brain normally would make readily but that is lost in the service of altruistic behavior.

Loss of information takes place in the constant neurobiological process by which we distinguish self from other. Put simply, fear in a high-stakes situation can evoke an ethical response by impairing the sharpness of this distinction, suddenly making it unclear who is at risk. Consider a person's action toward another. Before this action occurs, it is represented in the person's brain, as every act must be. Moreover, this act will have consequences for the other individual that the person can understand, foresee, and remember. Now comes the crucial step. The person blurs the difference between the other individual and himself. Instead of foreseeing the consequences of his act solely for the other individual, he sees them for himself, as well.

Let me give an extreme example, posed for absolute clarity. If someone is planning to knife another person in the stomach—with gruesome effects on the other person's guts and blood—he loses the difference between the other person's blood and guts and his own. He is less likely to carry through with the knifing he planned because he shares the other person's fear, indeed, experiences it in his own body. This is easy to explain, because the loss or impairment of any one of the many complex steps involved in the identification of other individuals would lead to the loss of information my theory supposes.

This loss of information—the person's blurring of the distinction between himself and the intended target of his action—is an essential element of my theory. How could it happen? The recognition of another person depends on a long series of electrophysiological and biochemical reactions to the stimuli particular to that person. These stimuli include seeing the other person's face, hearing his voice, feeling his touch, and smelling his personal odors. Every one of our senses works hard to identify that someone is that very person, there, not another person, and not ourselves. Reduction of the ability to make such discriminations in any of these sensory pathways will result in a blurring of the target's identity. He is less easy to discriminate from others and, in fact, from oneself.

This theory is not unreasonable. After all, explaining sensory discrimination is a daunting task, much harder than explaining how a computer works. Now, think of how easy it is to make a computer not work so efficiently. Improving the performance of a complex device (or biological process) is difficult. Degrading it—as my theory supposes—is ridiculously easy. Therefore, blurring the target's identity and thus making it more like your own is, in terms of brain mechanisms, reasonable to suppose. In fact, making a person's identity your own is essential to empathy.

The Role of Social Recognition

If my theory posits a loss of social distinction, it would be useful to understand how social recognition works. How do we recognize another person for who he is, as distinct from ourselves?

Scientists are beginning to piece together the molecular basis of social recognition through brain research on laboratory animals. Consider laboratory mice, which rely on smell to recognize others. Because virtually all social odors are signaled through forebrain pathways that lead to the amygdala, this brain region once again comes into play. Pheromones—odors that inform members of a species about other members of the same species—are recorded by specific parts of the amygdala. What then happens in the amygdala that is important for social recognition was studied by Jennifer Ferguson, Ph.D., and her collaborators Larry Young, Ph.D., and Tom Insel, M.D., at Emory University and by my laboratory at Rockefeller University. When influenced by sex hormones, larger amounts of the neuropeptide oxytocin meet larger concentrations of oxytocin receptors in the amygdala. Such a development is crucial, because Ferguson has shown that the administration of oxytocin to the amygdala fosters better social recognition. Young has also implicated a different neuropeptide, vasopressin, as important for fostering friendly social behaviors in laboratory animals.

Working together, oxytocin and vasopressin encourage experimental animals to act toward each other with positive, friendly behaviors. This emerging neurobiology for normal harmonious social interactions in animals, including humans, provides a positive set of mechanisms for ethical responses in a neutral or friendly situation. This adds to the mechanisms described earlier that apply to fearful situations. Even facing a high-stakes threatening situation, the great majority of humans will make an ethical response, because fear acts to do the job when the friendly hormones are not adequate to it.

Shared Fear, and Beyond

Usually, we have to recognize and remember differences between ourselves and others. For neuroscientists, my proposed explanation of an ethical decision by the would-be knifer is attractive because it involves only the loss of information, not its acquisition or storage. Learning and remembering complex information are difficult processes to understand. But the loss of information requires only the breakdown of any single part of the complex memory-storage processes, whether they are intricate biochemical adaptations, tricky synaptic growths, or precise temporal patterns of electrical activity. Eliminating or altering any one of the many mechanisms involved in social recognition or memory allows us to identify with the person toward whom we are about to act. Moreover, the specific mechanism that is affected could differ from person to person and from occasion to occasion.

In mechanistic terms, therefore, it is incredibly easy to achieve a sense of shared fate with another. In my example of the potential knifing, because of a blurring of identity—a loss of individuality—the knifer temporarily puts himself in the other person's place. Because that person would be afraid, so is he. He avoids an unethical act because of shared fear.

Is that all there is to explaining why we obey the Golden Rule when the chips are down? No, a more positive approach also exists. While I have emphasized an avoidance of the negative through shared fears, I have also invoked brain mechanisms, such as the action of the neuropeptides mentioned above, to explain positive "affiliative" behaviors toward others. That is a story for another article, but it would begin with the brain and behavior research in animals that asks how people recognize and tell each other apart, as sketched above. Friendly behavior begets compassion and sympathy—words derived from the Latin and Greek for "feeling with"—and we can envision a brain-based explanation for positive behaviors that also involves shared fates.

I have sought to explain behavior that follows an ethical principle when a just decision is to be made. But, as we know all too well, not everyone behaves ethically all the time. Why do some people find it harder to behave ethically than others? Why do some people have positive, friendly dispositions and others do not? Can neurobiology add to our understanding of what goes wrong when an ethical universal is violated, whether as a petty unkindness or a horrible act of aggression? These questions, too, await further exploration by scientists.

It may seem almost incredible that modern neurobiology has begun to explain the virtually universal ethical principle of treating others as we would wish to be treated. If this scientific explanation does provide an understanding of the underlying brain mechanisms, we should nevertheless remember that neuroscience is a new partner, not a replacement, for other intellectual approaches to the discipline of ethics. Certainly, the more we know about the mechanistic basis of empathy and ethical behaviors, the more effectively we can deal with situations in which we are saddened by their absence.

CHAPTER 4

Unshackling the Slaves of Obsession and Compulsion

A Brain Science Success Story

Judith L. Rapoport, M.D.,
and Gale Inoff-Germain

Judith L. Rapoport, M.D., is chief of the Child Psychiatry Branch, National Institute of Mental Health. She received the 2005 Edward M. Scholnick Prize in Neuroscience, as well as many other awards, for her research on the brain basis of obsessive-compulsive disorder and other research on childhood brain disorders. She is the author of the best-selling *The Boy Who Couldn't Stop Washing* (Signet paperback reissue, 1991). She can be reached at judy.rapaport@nih.gov.

Gale Inoff-Germain is a research psychologist at the National Institute of Mental Health. She specializes in developmental psychology and child psychiatry, with a special focus over the past decade on obsessive-compulsive disorder. She can be reached at ggermain@mail.nih.gov.

SAMUEL JOHNSON, one of the literary giants of the 18th century, leapt in and out of doorways before finally entering them and repeatedly touched buildings he passed while he walked the streets of London. Johnson suffered from what we now recognize as a severe case of obsessive-compulsive disorder (OCD) and complex motor tics. Among other famous figures who probably had OCD were the religious leaders John Bunyan, who was plagued by blasphemous thoughts damning God during his sermons, and Martin Luther, who continually feared that he had left out some important part of a prayer or sermon—an omission that would be a terrible sin. Both men had obsessions, that is, repetitive and unwanted thoughts that they could not control.

OCD is a common, often chronic illness that affects 2 to 3 percent of the population, according to many studies. People severely affected with OCD are imprisoned by sterile and frustrating self-imposed rituals (compulsions) and ruminations (obsessions). Some spend many hours every day washing themselves or checking something. For example, one woman never left home because of her compulsive need to check and re-check that the doors and windows of her house were locked. A teenage boy developed hypothermia and was taken to the hospital after his endless showering used up the household's hot water. Even when the water became cold, he could not leave the shower because he did not "feel clean enough." A successful newspaper reporter could no longer entertain anyone in her home because her inability to throw out newspapers and magazines had turned it into a virtual warehouse for old paper. She came to treatment only after her house was condemned as a fire hazard. These are examples of extreme OCD, but the quality of life for most people with the disorder is compromised by boring and endless preoccupations with unlikely dangers or fears of contamination.

Although OCD has been recognized for more than a hundred years, the past twenty-five years have seen an explosion of new information, new treatments, scientific excitement, and optimism among physicians. Once viewed as a quintessentially psychological disorder, OCD is now a model for how brain research can lead to understanding and treating a common, long-standing neuropsychiatric disorder. Inspired in part by this work, other anxiety disorders such as separation anxiety and specific phobias are increasingly being characterized in terms of specific brain pathways that do not function normally. Effective treatments are documented for most (although not all) adults and children with OCD, and today this disorder is seen as a classic success story of neuroscience and neuropharmacology. The principal challenges that remain are to find new means to help the significant minority of patients who do not respond to known treatments and to make all forms of treatment more widely available. Also an issue is the tendency of both researchers and physicians to label too many disorders "OCD" or "OCD spectrum," which helps neither patients nor the direction of research. But first, the OCD story.

OCD Basics

People with OCD are usually as sane and rational as the rest of us, but their very normalcy can make them want to keep their "crazy" symptoms a secret—one reason OCD was long believed to be rare. Even people with OCD who are in treatment for, say, depression, may not mention their OCD symptoms to their therapist. Who would want to reveal a compulsive need to count and recount the number of tiles on the floor, to tie and retie shoelaces to make them symmetrical and "just right," or to confess endlessly to nonexistent sins?

Although science now has a far better understanding of OCD, that understanding has not reached the lay public. When people (particularly children) first experience symptoms of OCD, they are bewildered by what they find themselves doing and thinking. One bright and imaginative boy wondered whether he was being used by Martians to carry out their orders and thought the Martians were looking for a sign that he had received their "message." He simply could not explain his symptoms (repeatedly entering and reentering each room) to himself any other way, so he was relieved to meet other children with similar behaviors, which scientists call motor rituals. Each new generation of patients needs to be educated that OCD is fairly common, related to brain abnormalities, and treatable.

Early theories about the causes of OCD focused on psychosocial influences, such as parenting styles, and the role of psychological defense mechanisms. Even traditional analysts, however, noted that psychoanalysis was seldom an effective treatment, and Sigmund Freud presciently maintained that OCD was a brain disorder.

In the 1970s, several independent studies launched a revolution in the understanding of OCD. First, a five-city epidemiologic study in the United States—a kind of psychiatric "census"—showed that OCD occurred in 2 to 3 percent of the population, or about 7 million people according to the population of the United States today. Second, both anecdotal evidence and pilot studies suggested that the new group of antidepressants known as serotonin reuptake inhibitors (SRIs) might be helpful not only for depression but also for OCD. This led to large groups of patients with OCD being sought for double-blind, placebo-controlled trials to make sure that the drugs actually worked on obsessions and compulsions, not just on depression.

For those of us involved in running the trials, the results were unforgettable. At first, the change in the patients we were studying was gradual, but by the end of the eight-week trials, it was clear that the drug that increased brain serotonin helped and that the placebo and other antidepressants that did not affect serotonin did not help. People who had had OCD for years described how they could now "shrug it off" or how the unwanted thoughts and impulses were somehow fading. We saw people reclaim their lives, go back to work. One man who had lived in isolation for years emerged and married his high school sweetheart. A young girl who had previously avoided playmates because of her contamination fears was able to stand in front of her sixth-grade science class and explain how she overcame her OCD.

Once researchers began seeing significant numbers of people with OCD, often along with their families, some striking patterns appeared. Although relatively rare cases of OCD are induced by trauma (for example, rape or other violent assault), most cases are not. About 20 percent of the patients had one or more family members with OCD, and about 50 percent of the adult patients said the disorder had started before they were 15 years old. That discovery has led to increased study of children with OCD. In a few important ways, childhood OCD differs clinically from the adult pattern. Childhood-onset OCD is more likely to run in families (that is, it may be more genetic), and it is more likely to be present together with motor tics (for example, excessive eye blinking or shrugging of the shoulders) or Tourette's syndrome (a combination of motor and vocal tics). A subgroup of children may develop OCD as a result of an autoimmune reaction to streptococcal infection.

The New Neurobiology of OCD

On the basis of the initial epidemiologic and pharmaceutical studies, and brain imaging studies that followed beginning in the 1980s, the brain circuitry of OCD has become one of the most clearly and meaningfully delineated brain dysfunction patterns known.

As scientists had seen, the first important class of drugs discovered to dramatically benefit people with OCD involved the neurotransmitter serotonin. Although many laypeople assume that drugs are developed from a biological understanding of a disorder, the converse is often true. For example, the effectiveness of antipsychotics was discovered accidentally, but subsequent studies of the action of these drugs led to the recognition that the dopamine system was important in understanding psychosis. The discovery that the serotonin reuptake inhibiting drug clomipramine, as well as the selective serotonin reuptake inhibitors (SSRIs) such as Prozac, were useful for OCD provided the clue that brain serotonin might not be normal in OCD. The study of the biology of OCD has also broadened to include other neurotransmitters and compounds, such as those that affect dopamine and, more recently, glutamate.

Neuroimaging techniques—including computer-assisted tomography (CT), positron emission tomography (PET), and, most recently, functional magnetic resonance imaging (fMRI)—were critical to understanding OCD. Studies have shown smaller brain volumes and abnormal patterns of brain activation in several regions, including the orbital frontal cortex and the basal ganglia, in people with OCD (Figure 1). These dis-

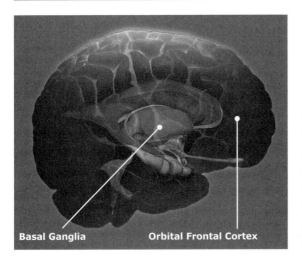

Basal Ganglia   **Orbital Frontal Cortex**

Figure 1   This composite 3D MRI and CT scan image shows two parts of the brain that have been shown to be abnormal in people with OCD: the orbitofrontal cortex and, deep within the brain, the basal ganglia. *Credit: © ISM / Phototake*

coveries were supplemented by magnetic resonance spectroscopy studies suggesting that OCD may involve brain abnormalities in the neurotransmitter glutamate.

Hypotheses about possible abnormalities in the part of the brain called the basal ganglia in people with OCD have come not only from noting the frequent connection between OCD and motor or vocal tics but also from case reports of sudden onset of OCD after local damage to this brain region. Literally, basal ganglia means "lower nerve knots," and the ganglia are indeed buried deep within the brain. Animal studies have shown this part of the brain to be important for planning complicated motor movements and for making decisions about these movements. These reports of OCD associated with basal ganglia damage were followed by the fascinating observation that several neurological problems thought to be related to basal ganglia damage (such as Sydenham's chorea, tics, or Tourette's syndrome) are also associated with OCD. That is, children and adults with basal ganglia disorders are likely to also have OCD. The observation that Sydenham's chorea is associated with OCD provides support for an autoimmune response model of OCD, because, in this chorea, an autoimmune response to streptococcal infection leads to antibodies that attack selective regions of the patient's brain.

Brain imaging studies have also shown that a part of the cerebral cortex, the orbital frontal cortex, appears to be overactive in people with OCD. This region, too, is involved in complex planning and is also connected to systems in the temporal lobe that modulate emotions. Thus, many clinical aspects of OCD, such as the overfocus on planning and extreme anxiety, are now coming together with the neurobiological research.

A Far from Uniform Disorder

Although commonalities exist, several subgroups of people with OCD have been definitively or tentatively identified. For example, treating abnormalities in the dopamine system seems important in people with both OCD and a tic disorder, making up one possible subgroup. This subgroup appears to overlap considerably with childhood-onset OCD. Determining the characteristics that are common to all people with OCD, compared with those that are distinctive to subgroups, will help scientists better understand what causes OCD and will also allow more individualized treatment planning.

Another possible subgroup, which appears to involve an autoimmune dysfunction, was defined in children and labeled with the acronym PANDAS, for pediatric autoimmune neuropsychiatric disorders associated with streptococcal infection. The PANDAS syndrome is diagnosed by specific criteria, including the presence of OCD or a tic disorder and two or more episodes with abrupt onset or sudden worsening of symptoms after streptococcal infection.

Some controversy has long been part of the scientific debate about the role of immune processes in neuropsychiatric disorders. For example, non-steroidal anti-inflammatory drugs appear to be of some benefit to people with Alzheimer's disease, and studies of the brains of people who died of the disease have found indications of inflammation. Consequently, inflammation is suspected to play a role in causing Alzheimer's, although this has not yet been proved. Because both streptococcal infection and obsessive-compulsive symptoms are commonly found in children, it is difficult to be sure that the streptococcal infection is actually involved in causing the OCD. Figuring out cause and effect is also complicated because, in general, OCD and tics may come and go. But support for the role of antibodies in OCD and similar disorders comes from a variety of research, including initial studies of treatments based on an autoimmune approach that we discuss later.

The Genetics Behind OCD

Evidence from family and twin studies points to a genetic component of OCD, especially when it begins in childhood. Family studies let scientists see what disorders appear together in families. A key issue in these genetic studies is deciding what to count as OCD. Some family members may have tics, but no OCD symptoms. Other family members may have trichotillomania, a disorder involving hair pulling that is sometimes viewed as related to OCD. Studying entire families, therefore, helps researchers see what might be part of a whole "OCD spectrum," and looking at the genetic information using these broad or spectrum definitions as well as narrow ones (just classic OCD) is essential in modern genetic studies.

An outpouring of technological advances in genetic research has begun to transform the study of OCD. Newer methods will soon enable even faster and less expensive acquisition of genetic information, but promising results are already emerging. Candidate genes, selected, so far, on the basis of which drugs work best for OCD and on the brain circuitry discovered through imaging studies, are being examined to see whether they are associated with OCD. Studies that compare people with OCD and those without the disorder focus on whether particular chromosomal markers are more frequent in the first group. Probably most important is that two studies of families with multiple members who have OCD have pointed toward the same region on chromosome 9.

Another kind of research called cytogenetic study (involving the study of physical duplication, deletion, or disruption of chromosomes) has also yielded an intriguing result. In several such studies, a chromosome deletion known as the 22q11 deletion (which results in a disorder called velo-cardial facial syndrome) was found to be associated with OCD, particularly in children.

As yet, these advances in genetics are too preliminary to benefit people with OCD or their families, but scientists know that at a certain point an understanding of genetics leads to, or combines with, an understanding of physiology to make huge gains possible. In schizophrenia research, for example, studies of the gene COMT (which regulates neurotransmitter metabolism) have shown different brain responses in people with schizophrenia who have the risk form of the gene (the "risk" allele) when they perform certain cognitive tests and in normal control subjects. Other studies of risk genes involved in depression have shown different brain patterns in response to faces expressing emotion in people with the risk form of those genes.

Treatments, Proven and Potential

Cognitive Behavior Therapy

Pharmaceuticals are only one approach to treating OCD. Cognitive behavior therapy (CBT) was shown to be effective for both children and adults with OCD, and for many people it will be the first and only treatment that they will need. The most effective part of behavior therapy is usually the repeated and voluntary exposure to something that triggers the OCD, for example, touching something "contaminated" and then not washing for a certain period of time. Each symptom is addressed in turn, and it usually takes several weeks to see results. Nevertheless, CBT has a great advantage over drugs: it continues to work when the therapy has stopped, but drugs typically are effective only while they are taken.

Many kinds of behavior therapy exist. For example, the commercial Weight Watchers plan uses behavior therapy, and the physiological training method called biofeedback is also a kind of behavior therapy. But these techniques are different and will not help people with OCD. For these people, it is important not only to have a trained behavior therapist but also one experienced with OCD and with what is called exposure with response prevention (ERP). CBT/ERP has been studied for decades, but only recently have the studies begun to include large sample groups of subjects and appropriate control subjects (comparison groups) so that firm conclusions may be drawn. The consensus, though, is that, if CBT/ERP is available, it should be tried first, before drugs. Moreover, recent studies in both adults and children show that the combination of medication and CBT can be more effective than either approach used alone.

SRIs

Initial research on treating OCD focused on the SRIs, drugs that increase serotonin availability at the nerve endings. What is particularly helpful in treatment is that, although one SRI or SSRI may not be sufficiently effective or may produce unpleasant side effects for a particular patient, a different drug may be both effective and without unwanted side effects for that person. Drugs approved for the treatment of OCD are listed in Figure 2.

Figure 2   Drugs Approved for Treatment of OCD:

clomipramine (Anafranil)

fluoxetine (Prozac)

fluvoxamine (Luvox)

paroxetine (Paxil)

sertraline (Zoloft)

Note: **Bold** text indicates those drugs approved by the Food and Drug Administration for children and adults.

Augmenting Drugs

Complete remission of OCD symptoms in response to a single drug (typically an SSRI) is rare. Physicians will therefore try several different drugs with patients and experiment with dosage. But after that, a combination of an SSRI with another drug may be used for those who are still severely affected (sometimes called "treatment resistant" or "treatment refractory"). These secondary drugs are used to "potentiate" or "augment" the effects of the first.

Drugs that affect the dopamine system were among the first to be shown (in a double-blind, placebo-controlled trial) to be effective as augmenting agents for OCD. Many of those currently used are atypical antipsychotics, such as risperidone or olanzapine. But their long-term effects when used in OCD are still largely unexplored, and information describing use with children and adolescents is limited. These drugs may also cause weight gain and have other negative effects on the endocrine system, so patients need to be followed carefully by a physician.

Autoimmune Treatments for PANDAS

A notable treatment success is based on the autoimmune model of PANDAS. For example, a double-blind, placebo-controlled study of intravenous immunoglobulin (IVIG) showed improvement for the patients who received the IVIG and not for the placebo group. An open trial (that is, a trial without a placebo control) of plasmapheresis (plasma exchange) also looked encouraging. For several reasons, these "immunomodulatory" treatments are not currently being considered for general use, but they are prompting sufficient interest to encourage continued research.

The only published double-blind study of penicillin treatment with a placebo control group did not show an improvement in symptoms from the antibiotic. Despite this finding, some physicians are using penicillin prophylaxis for children whose OCD appears to spike repeatedly in relation to streptococcal infections. Clearly, this study should be replicated, perhaps with a more carefully chosen group of subjects.

New Drug Treatments

Although the same particular biological processes and pathways may be disturbed in most people with OCD, it is not clear which biological systems are the main sources of the problem. This complexity compounds the difficulty of developing treatments, but new approaches are continually being explored on the basis of increased knowledge about OCD.

For example, although it is still early in the research, some open trials and limited results suggest that opiates such as oral morphine sulphate or glutamate antagonists (drugs such as riluzole that block or limit effects on glutamate) may be helpful. The positive results of using these drugs in people with OCD support the suspicion that signaling pathways in the brain for neurotransmitters other than serotonin may be involved in some kinds of OCD.

New Nondrug Treatments

Surgical efforts to treat psychiatric disorders during the first half of the 20th century appropriately resulted in a public skepticism that persists to this day, but over the past decade neurosurgery has been revolutionized. Although still used only for the most severely incapacitated and "intractable" patients, and only after approval by formal medical and ethics boards, surgical treatments use dramatically improved techniques. One such technique is radiosurgery, which involves lesions produced by controlled cross-fired irradiation at localized sites in the brain. Still, this and other neurosurgical procedures must be considered experimental, at this stage, and are unlikely to be studied in children.

Other brain-based approaches being explored in an attempt to help those who are unresponsive to available treatments are at the same time clarifying how the brain works and how various treatments achieve their results. The most noninvasive technique of this type is transcranial magnetic stimulation (TMS). TMS is used with awake patients, requires no anesthesia, and produces direct cortical brain stimulation by creating a transient magnetic field that induces electric currents in the brain. So far, the technique is not particularly promising as therapy for OCD, but TMS has yielded information about OCD brain circuitry and may have a role in guiding or screening for more-invasive procedures.

Another new area is deep brain stimulation (DBS), which appears to produce some of the outcomes previously obtained by neurosurgeons when they made lesions in the brain. Initial evidence of its effectiveness came in treatment of movement disorders, particularly tremors and Parkinson's disease. DBS is a flexible and reversible technique that involves high-frequency stimulation of neurons at sites important for movement disorders. Although DBS is still a new tool, some preliminary evidence for its effectiveness exists, and a few severely ill patients with OCD are being studied.

Does an OCD Spectrum Exist?

Because similar therapies appear to be effective in treating them, other disorders characterized by repetitive and overfocused behaviors have been suggested as part of an OCD family, or spectrum, of disorders. Included are eating disorders such as anorexia nervosa, trichotillomania, body dysmorphic disorder (a preoccupation with an imagined physical defect in appearance or a vastly exaggerated concern about a minimal defect), autism, and Asperger's syndrome. For some of these disorders, however, including trichotillomania and anorexia, the SRI drugs are not particularly beneficial. For others, such as autism, some help comes from these drugs, but the results are usually not dramatic.

For now, this particular spectrum grouping does not seem to be useful in terms of clinical treatment. The drugs used for OCD seem less helpful for these other disorders, and behavior therapy appears more difficult. Many psychiatric disorders, after all, involve excessive activity or overfocus in one sphere of life, and the proposed OCD grouping potentially could become ever larger without providing more useful information. Until a specific brain abnormality or a clear and salient risk gene is identified for OCD, scientists may have little to gain by identifying such a spectrum. This lesson can be applied more widely. Broad definitions of the disorder were helpful for family research studies on schizotypal personality and other schizophrenia spectrum disorders but have not helped treatment.

Caveats to the Success Story

Most success stories have limitations, and OCD is no exception. For at least 25 percent of people with OCD, treatment is not effective, even in those people who have tried both behavior therapy and drugs. People with severe or debilitating OCD for whom drugs are effective face the expense of lifelong treatment. Some types of OCD (for example, severe hoarding) are particularly difficult to treat. These are both problems and potential clues to new understanding.

Answers will come from additional research. For example, researchers still need to explore the relation of OCD to neuroendocrine changes. OCD exacerbations often occur in pregnancy, but scientists do not know why, nor do they understand whether a relation to diabetes mellitus exists, as has been suggested. Also, continuation of sophisticated magnetic resonance spectroscopy studies will undoubtedly lead to investigating processes in the brain such as the glutamate system.

As is the case with other chronic diseases, the cost of appropriate treatment may be beyond the reach of many people with OCD. Another concern is that outside large cities the availability of therapists trained in CBT/ERP is severely limited. These daunting obstacles continue to limit the extent to which OCD can be treated, affecting patients, family members, and the health-care professionals who put forth so much effort and determination in meeting the challenges of OCD. The results are considerable human and economic costs to these many patients, their families, and their communities.

Despite all the successes so far in research and treatment, the fruits of this neuroscience success story are not available to all who need them so desperately. It is particularly frustrating to have this problem now, when the research has never been so exciting and so ready to be translated into proven new treatments. An important first step would be parity in reimbursement for all psychiatric disorders, comparable to that for diseases labeled physical.

The OCD story is perhaps the most dramatic example yet of how a once purely "psychological" disorder is now understood in largely biological terms. This success also shows the importance of interdisciplinary research, because it took a combination of epidemiologic, pharmacologic, brain imaging, neurosurgical, and immunologic studies to work out the OCD puzzle. Scientists must find a way to train even more people who are knowledgeable about both clinical psychiatric syndromes and these synergistic fields of basic research so that what they have learned in understanding OCD will truly be a model for what modern neuroscience can achieve.

IF YOU OR SOMEONE YOU ARE TRYING TO HELP may have OCD, an important step is professional diagnosis of the disorder. Psychology or psychiatry departments at universities and medical schools have lists of physicians, social workers, and behavior therapists. The Obsessive Compulsive Foundation, a nonprofit national organization, has a Web site (www.ocfoundation.org) that offers up-to-date information and resources, including a referral list of mental health professionals for your geographic area and a low-cost newsletter. The National Association of Cognitive-Behavioral Therapists (www.nacbt.org) is another useful resource. These organizations can help you find qualified professionals, get reliable information, and connect with support groups.

CHAPTER 5

Are We in the Dark About Sleepwalking's Dangers?

Shelly R. Gunn, M.D., Ph.D., *and* W. Stewart Gunn

Shelly R. Gunn, M.D., Ph.D., received both her M.D. and her Ph.D. in neurogenetics in 2002 from the University of Texas Health Science Center at San Antonio (UTHSC-SA), where she has also been an instructor of medical neuroscience. She is currently completing a residency in clinical pathology at UTHSCSA. Her work focuses on the use of genome-scanning techniques for the clinical analysis of congenital abnormalities and blood cell cancers. She can be reached at gunn@uthscsa.edu.

W. Stewart Gunn is a freelance writer and a history major at Texas Christian University in Fort Worth, Texas. He can be reached at lormaster3@aol.com.

SLEEPWALKING GENERALLY OCCURS IN THE DARK and has remained there, both literally and figuratively, for centuries. The image that comes most readily to mind is a cartoon person, amiably and aimlessly wandering the hallway with arms outstretched and eyes closed. But sleepwalking is not funny; it is a sleep disorder known to specialists as somnambulism. Many adult sleepwalkers, with eyes open, perform purposeful acts such as eating half a bag of chips and putting the rest in the microwave, taking all their shoes from the closet and lining them up on the windowsill, rearranging furniture, or climbing out a window in the middle of the night—activities that are essentially benign when a person is conscious but that, when they occur during somnambulism, are potentially dangerous to the sleepwalker or other people. More frighteningly, increasing numbers of "sleepdriving" cases are being reported in which somnambulists get in their cars and drive sometimes long distances, disregarding lanes, stoplights, and stationary objects, and, after waking up, have no memory of what they did.

Although these nocturnal wanderings may seem extremely odd to non-sleepwalkers, such mechanistic and automatic activities are part of the spectrum of behavior associated with somnambulism, which is estimated to affect close to 2 percent of the adult population worldwide. Sleepwalking and other sleep disorders appear to be on the rise in our demanding and fast-paced society, in which getting a good night's sleep is increasingly difficult. Many people resort to prescription (or nonprescription) drugs to induce sleep, but sometimes this only compounds the problem. Sleep deprivation, especially in combination with drugs and alcohol, is known to induce sleepwalking in some people, and behavior while sleepwalking is extremely unpredictable, particularly in a new environment. Any adult with a tendency to sleepwalk has the potential to experience an accident and can be at risk of real injury. But, until recently, published reports of injuries as a result of sleepwalking were rare, and somnambulism and other sleep disorders are frequently overlooked in the medical school curriculum.

Although I am a teacher of medical neuroscience, the dangers of sleepwalking would probably never have come to my attention had my son Stewart (who joins me in writing this article) not sleepwalked out a second-story window into an alley, sustaining serious injuries, on the night he arrived for a British Studies program at St. John's College in Oxford, England. His potentially fatal experience with sleepwalking demanded a reexamination of this overlooked topic and raised many questions during his convalescence. Had other people been seriously injured while sleepwalking? If so, were these random and rare events, or had we encountered the tip of an unexplored iceberg?

By searching the medical literature and interviewing other sleepwalkers, we found that sleepwalking accidents and injuries, more common than usually believed, are a definite health hazard for both the sleepwalker and other people. But such accidents are not well known, because both the general public and physicians are uninformed about somnambulism.

In this article, we explore current theories about both the causes and the management of adult sleepwalking, while seeking to increase awareness of its hidden dangers. Sleep medicine needs to be an integral part of the medical school curriculum, and physicians as well as the general public should be aware that, unlike sleepwalking in children, somnambulism in adults is a potentially dangerous disorder. Both treatment of the disorder—when possible—and prevention of accidents are of paramount importance for the sleepwalker and for unsuspecting people who may find themselves in the sleepwalker's path.

Experiences of Sleepwalkers

Stewart arrived in Oxford, sleep-deprived after the long trip from Texas, and checked into his second-floor dormitory at St. John's College on a warm July day. The wide-open windows had no screens and were surrounded by scaffolding. Even though he had been awake for more than thirty hours, he chose to postpone sleep until after dinner in order to adjust to British time. He fell asleep easily, but when he awoke about 2:00 a.m., he was lying facedown on a cobblestone street that he did not recognize. He had absolutely no recall of having left his dorm room, walking through several doorways, and stepping out a window onto the scaffolding from which he must have fallen into the alley. After trying unsuccessfully to lift himself off the cobblestone street, he dragged himself toward what appeared to be a road and was discovered by the British police. He had fractured his spine and right wrist in the accident, but thankfully had no permanent neurological damage.

At the same time that Stewart was recovering at the John Radcliffe Hospital, CNN News reported the story of a London teenaged girl who was rescued from the top of a crane, which she had climbed while sleepwalking and then gone back to sleep on the support beam—fortunately catching the attention of a pedestrian, who notified the police. The hospital staff caring for Stewart found this amusing and suggested that the two young sleepwalkers get to know each other. Although Stewart and the girl never met, their similar stories motivated us to start looking further into the prevalence and possible causes of adult sleepwalking.

We did not have to look far for stories. The paramedic who helped take Stewart to Gatwick Airport had an adult sister who had injured herself while sleepwalking, and we encountered a woman on the flight home whose son had repeatedly sleepwalked onto a balcony. An acquaintance called to relate how her son not only regularly disassembled bedside lamps while sleepwalking but was recently found, in his pajamas, pumping gas after sleepdriving to the gas station. In 2003, several cases involving mysterious nighttime accidents, some of which were fatal and initially ruled as suicide, were reported in the *Journal of Forensic Science* by Mark Mahawald, M.D., director of the Minnesota Regional Sleep Disorders Center. These deaths, referred to as "parasomnia pseudosuicides," were later attributed to complex motor behaviors that can take place during sleepwalking, such as running, climbing, or jumping. One involved a twenty-one-year-old college student who was hit by a semitrailer after he ran onto a highway at 4:30 a.m., clad only in his boxer shorts. He had no history of drug abuse or depression, but he and several family members had a history of frequent, complex sleepwalking. A formal review of his case requested by his family resulted in a recommendation by the medical examiner that the cause of death be changed from "suicide" to "accidental death due to sleepwalking."

Other cases described by Mahawald involved falls from balconies, defenestration (jumping from windows), and self-inflicted gunshot wounds by people with a past history of complex sleepwalking behavior and no history of depression. Distinguishing between accidental death and suicide has profound religious, societal, and insurance implications, of course, and many of the families of these victims requested that these pseudo-suicides be reevaluated as accidental death as a result of a sleep disorder. Some cases of homicidal behavior during sleepwalking have also been reported. The legal defense in these cases has usually been to claim that the action was a "non-insane automatism," meaning that the brain's motor system was fully aroused but consciousness was clouded. In all of these types of cases, a correct diagnosis or verdict can be made only if the family, police, and medical examiners are willing to consider alternative scenarios.

Prescription Drugs as a Risk Factor

Early in 2006, a surge of news reports described complex sleepwalking behaviors that involved binge eating, violent outbursts, and sleep-driving in people who took the medication Ambien (zolpidem).

Ambien, the best-selling prescription sleeping pill in the United States, is a popular alternative to the more traditional benzodiazepines because of its general lack of serious side effects. Since it was introduced in the early 1990s, mild side effects such as nausea, dizziness, and nightmares have been reported in people taking the drug as prescribed. But in 1994 and 1995, the first two cases of sleepwalking attributed to Ambien appeared in the medical literature, followed by an additional six cases during the next ten years. The most recent case, reported in the *Archives of Physical Medicine* in June 2005, involved a middle-aged man with no history of sleepwalking or previous Ambien use who began sleepwalking after taking Ambien while hospitalized for hip surgery. His somnambulism stopped as soon as the drug was discontinued.

A 2006 article in the *New York Times* reported that Ambien is one of the top ten drugs identified in the blood of impaired drivers, and, in

Wisconsin alone, Ambien was identified in 187 drivers arrested from 1999 to 2004. In a presentation for the American Academy of Forensic Sciences, Laura J. Liddicott and Patrick Harding of the Wisconsin State Laboratory discussed six cases now set for trial. They reported that all of the drivers tested negative for ethanol and other drugs but had serum levels of Ambien well above the therapeutic range of 29 to 272 ng/mL (nanograms per milliliter). Each of the drivers displayed extremely bizarre behavior, such as wide deviations from the marked lanes and near collisions with stationary objects, and all of them, when stopped by police, appeared confused, disoriented, and somnambulant, with no memory of what they had just done. A class-action lawsuit has been filed by New York attorney Susan Chana Lask against Sanofi-Aventis, the company that makes Ambien, on behalf of anyone nationwide who has experienced such side effects as sleepwalking, sleepeating, sleepdriving, or memory loss while taking Ambien.

The official position of Sanofi-Aventis is that "the safety profile of Ambien is well established and reported in the Ambien Prescribing Information approved by the U.S. Food and Drug Administration. Sanofi-Aventis regularly conducts thorough analyses and has not observed any significant change in that safety profile. The information currently contained in the U.S. Prescribing Information remains accurate: sleepwalking (somnambulism) is a possible rare adverse event."

During the past ten years, a handful of case reports were published that describe sleepdriving in non-Ambien users. These reports described behavior not unlike that seen in Ambien sleepwalkers: long-distance driving and bizarre behavior, followed by complete amnesia for the event. The unifying question that needs to be addressed is, What is happening in the brain to cause such a disassociation between being awake and being asleep?

Inside the Sleepwalking Brain

In their classic overview of normal human sleep found in *Principles and Practice of Sleep Medicine* (4th edition, 2005), Mary Carskadon,

Ph.D., professor of psychiatry at Brown University, and William Dement, M.D., Ph.D., founder of Stanford University's Sleep Disorders Clinic, describe sleep as "a reversible behavioral state of perceptual disengagement from and unresponsiveness to the environment." In the *Oxford Dictionary*, sleep is defined as "a condition of body and mind such as that which normally recurs for several hours every night, in which the nervous system is inactive, the eyes closed, the postural muscles relaxed, and consciousness practically suspended." By both these definitions, a sleeping person is basically out cold.

In sleepwalkers, however, the states of being awake and being asleep are not mutually exclusive; instead, they occur simultaneously. Parts of the brain are aroused, the eyes are open, and postural muscles are tensed and active, while clear, lucid consciousness remains suspended. Sleepwalkers, who are awake and asleep at the same time, have been described for centuries. Think of Shakespeare's Lady Macbeth: "You see, her eyes are open . . . but their sense is shut."

More analytical descriptions of human sleep became possible at the beginning of the twentieth century, with the development of the electroencephalogram (EEG), a device that measures and records the electrical activity of the brain by using flat metal discs placed on the surface of the head. The electrodes are connected by wires to an amplifier and recording machine that convert the electrical signals from the brain into pen-and-paper tracings that resemble waves. Further technological advances have increased the ability of researchers and physicians to analyze sleep and its disorders by using all-night monitoring of not only the brain but also eye and leg movements, respiration, and heart rate through a comprehensive diagnostic test known as polysomnography, conducted in a sleep laboratory.

On the basis of EEG wave patterns, sleep can be broadly divided into rapid eye movement (REM) and non–rapid eye movement (nREM). REM sleep is associated with dreaming and with high-frequency, low-voltage brain waves, whereas nREM sleep is characterized by low-frequency, high-amplitude waves known as slow-wave sleep (SWS). nREM sleep is further divided into stages I to IV, with stages III and IV representing the deepest sleep. A normal sleeping adult will repeatedly alternate between nREM sleep and REM sleep, with each total cycle lasting on average about ninety minutes, and SWS dominating the first third of the night. A person who sleeps for eight hours will progress through four or five cycles during one night's sleep, with REM episodes becoming longer during the course of the night.

During REM sleep, a complete and dramatic loss of muscle tone occurs. This loss is protective, because it prevents the sleeper from acting out dreams. However, sleepwalking generally occurs during stages III and IV, possibly as a result of an incomplete transition from SWS back into REM sleep, and dreams are not commonly associated with these nREM stages. What, therefore, causes sleepwalkers to leave their beds and wander into the night?

Finding the answer to this question has been the goal of much research. Currently, several theories exist about what precipitates sleepwalking, although the underlying pathophysiology is still not well understood. An August 2000 research letter to the British medical journal *Lancet* described the results of an experiment on a sleeping subject by using a brain imaging technique called single photon emission computed tomography (SPECT), which showed how continued deactivation of the prefrontal cortex of the brain, in combination with abnormal activation of the cingulate cortex and the thalamus, may lead to the dissociation between "body sleep" and "mind sleep" characteristic of somnambulism. In addition, twin studies have shown a possible genetic association, and in 2003 a specific genetic marker, called the HLA subtype (DQB1*0501), was implicated in sleep-associated motor disorders by Michel Lecendreux, M.D., of the Robert Dobre Hospital in Paris, France.

nREM Sleep Instability

Sleep researchers have found that sleepwalking occurs against a background of nREM sleep instability characterized by a particular kind of high-voltage brain wave called hypersynchronous slow delta (HSD). These HSD waves were first described in a May 1965 article in the journal

Science describing a University of California, Los Angeles (UCLA) study of a group of sleepwalkers who underwent all-night EEG recording using special techniques that allowed them to get up and move around. HSD waves were recorded both before and during sleepwalking episodes in the UCLA study, and subsequent studies have consistently identified HSD waves associated with sleepwalking. However, the significance and specificity of these waves for sleepwalking have been questioned in subsequent research, because HSD waves have also been identified in the EEGs of people with other sleep disorders, as well as in normal sleep.

According to Christian Guilleminault, M.D., Ph.D., of the Stanford University Sleep Disorders Clinic, HSD waves can be seen normally in the nightly cyclical passage from stable to unstable sleep as the billions of neurons in the human brain are recruited during early stages of nREM sleep into the orderly SWS rhythms of stages III and IV. However, when the HSD wave pattern persists throughout stages III and IV, it is indicative of an interruption in the normal progression of nREM sleep. This disturbance can be interpreted by sleep researchers using a technique called cyclic alternating pattern (CAP) analysis.

CAP analysis is particularly valuable in evaluating nREM disorders of arousal, because these disorders are not generally accompanied by significant changes in the brain processes that can be detected by all-night sleep monitoring using EEG and the other tools of polysomnography. CAP provides a measure of the microstructure of brain waves in sleep instability through analysis of sequences of EEG patterns. If the CAP rate is indicative of abnormal sleep, then it is imperative to search for the instability's cause, which is generally a subtle associated sleep disorder. In a January 2006 article in the journal *Sleep Medicine,* Guilleminault stated that he did not find a "pure" sleepwalker in the most recent 100 cases he had studied and that the identification of the underlying cause of sleep instability often led to treatment and elimination of sleepwalking in his patients.

What Causes Sleep Instability?

In chronic sleepwalkers, respiratory syndromes are the most frequently diagnosed accompanying disorders. As a result of the close relation between abnormal retention of carbon dioxide in the blood (a condition known as hypercapnia) and activation of neurons within the brain stem that control sleep and waking, an inability to breathe normally affects neural control of the progression through sleep. Specific respiratory syndromes, including upper airway resistance syndrome, mild obstructive sleep apnea, and sleep-disordered breathing, have been diagnosed as underlying causes of sleep instability. Through CAP analysis of chronic sleepwalkers, researchers have learned that the basic nREM instability accompanying the breathing disorder is present even on nights when no sleepwalking occurs. But the instability almost always completely vanishes when the respiratory problem is successfully treated, usually through the delivery of air to the upper respiratory tract through a specially designed mask (continuous positive airway pressure, or CPAP) or through surgery.

The Stanford University study of chronic sleepwalkers reported by Guilleminault found that sleepwalking was much more likely to be eradicated in patients with treatable respiratory disorders, so it is important to seek an underlying cause of sleep instability for chronic sleepwalkers. The rare cases of "pure" sleepwalking, which appear to have no associated disorder, may represent a subgroup of sleepwalkers in whom nREM sleep instability is the result of genetic factors. Benzodiazepines, the most commonly prescribed drugs for sleepwalking, are only partially effective in eliminating sleepwalking in these patients, so attention must be focused on maintaining a safe sleeping environment to prevent accidents.

Living Safely with Adult Somnambulism

When sleepwalking behavior persists or reemerges in adulthood, it is no longer a relatively benign disorder of childhood, even though the same underlying nREM sleep instability is present at all ages. Occasional injuries have been reported in childhood sleepwalkers, but by the time a child is about twelve years old, when the central nervous system matures, episodes usually disappear—before most children are able to drive or have access to alcohol and guns. The most common automatic, unconscious behavior in young children who sleepwalk is to seek their parents, which is what Stewart did during his childhood sleepwalking episodes.

For adults in whom sleepwalking has become chronic or dangerous, it is thus imperative to address both issues of safety and the eventual elimination of the behavior. Since repeated episodes of somnambulism indicate an underlying nREM sleep instability, physicians must try to identify any associated sleep disorder that could be causing abnormal progression through the stages of sleep. But even if, as is most often the case, a sleep-related breathing disorder is identified, treatment such as CPAP or surgery is not instantly successful, and the inherent dangers of sleepwalking will persist until the cause of nREM instability is completely eliminated. Therefore, safety remains of paramount importance in managing chronic sleepwalking even after diagnosing and starting to treat the underlying disorder. In some sleepwalkers, no treatable cause will be found; for them, attention to sleep practices and safeguarding the environment are lifetime challenges.

The real key to sleepwalking safety is knowledge—knowledge of whether a person is a sleepwalker and awareness of the conditions or drugs that increase the possibility of a sleepwalking incident. All prescription sleep medications should be taken exactly as directed. Ambien may not be a good choice for someone with a history of sleepwalking. If taken with alcohol, Ambien has the potential to induce sleepwalking even in people with no previous history of the disorder. According to Laura Liddicoat, the forensic toxicologist who is investigating the Wisconsin sleepdriving incidents, the only tolerable blood alcohol level for someone who is taking Ambien is 0.0 percent.

Sleep deprivation is also known to trigger sleepwalking in susceptible persons, possibly as a result of the extremely deep nREM sleep, known as rebound or recovery sleep, that often occurs after long periods without sleep. Stewart had been awake for more than 35 hours when he finally fell asleep in his Oxford dormitory room, an amount of sleep deprivation that sleep laboratory studies have shown is sufficient to increase the frequency and complexity of somnambulistic episodes during recovery sleep. In some laboratories, artificially inducing sleepwalking by sleep deprivation has been used as a tool in the diagnosis of somnambulism. Known sleepwalkers should therefore do everything possible not to become sleep-deprived, particularly when they cross time zones and sleep the first night in a new environment—all factors in Stewart's Oxford accident. In addition, neither alcohol nor sleep drugs such as Ambien should be taken under these conditions.

Both at home and when traveling, safeguarding the environment should be a top priority for sleepwalkers. Appropriate precautions include choosing lower bunks and ground-floor rooms and bolting shut doors and windows, possibly with a chair placed in front of them (after first locating and not blocking the fire exits). Beds should be pushed to the wall, and a sleeping partner should sleep on the outside, so the sleepwalker would have to climb over the partner to get out of bed and wander into the night. If possible, bedroom and outer doors should be equipped with alarms and buzzers that are loud enough to awaken the sleepwalker or the family, particularly when traveling by boat because of the risk of falling overboard. Power tools or guns should be stored in locked cabinets with combinations or key entry not amenable to being unlocked in an unconscious state. A sleepwalker should never be allowed to drive while somnambulant, and car keys as well as the car should be made inaccessible at night if there is any tendency to sleepdrive. Contrary to what most people think, it is not dangerous to awaken a sleepwalker, and he or she will probably thank you the next morning upon waking up safely in bed.

For the most part, sleepwalkers have earned their amiable reputation as, in the words of Shakespeare, just "merry wanderers of the night." Whether conducting imaginary orchestras, climbing trees, or taking walks outside clad only in pajamas, they give us a glimpse of the incredible intricacy and complexity of a human brain that is capable of being awake and asleep at the same time. That most people successfully make the journey through the many stages of sleep several times each night is a testimony to the ability of billions of neurons to synchronize themselves into the fundamental biological process required by all organisms—the need to rest. In somnambulism this process has somehow been subverted, but, with a growing public awareness about the hidden dangers of sleepwalking and increasing coverage of sleep medicine in neuroscience textbooks and medical school curricula, we hope that the world is becoming a safer place for all nighttime wanderers.

CHAPTER 6

Knowing Sin *Making Sure*
Good Science Doesn't Go Bad

Henry T. Greely, J.D.

Henry T. Greely, J.D., is the Deane F. and Kate Edelman Johnson Professor of Law at Stanford University, where he directs the Center for Law and the Biosciences and the Program on Stem Cells in Society. Professor Greely has written widely on the legal and ethical implications of biological advances. He serves as chair of the California Advisory Committee on Human Stem Cell Research and chair of the steering committee of the Stanford Center for Biomedical Ethics. He can be reached at hgreely@stanford.edu.

Despite the vision and the farseeing wisdom of our wartime heads of state, the physicists felt a peculiarly intimate responsibility for suggesting, for supporting, and in the end, in large measure, for achieving, the realization of atomic weapons. Nor can we forget that these weapons, as they were in fact used, dramatized so mercilessly the inhumanity and evil of modern war. In some sort of crude sense which no vulgarity, no humor, no overstatement can quite extinguish, the physicists have known sin; and this is a knowledge which they cannot lose.[1]

—J. ROBERT OPPENHEIMER (1947)

LIKE ALL TOOLS, scientific advances may be used for good or for ill. As our knowledge about the human brain increases, we will certainly use that knowledge to relieve human suffering in profound and wonderful ways. But the vast promise of the science should not blind us to the possibilities of its misuse. I believe those involved in human neuroscience need to pay attention to the risks that come with the science and to accept the duty to minimize any harm it could cause.

The Dark Side of Science—Some Historical Examples

Nuclear weapons were first detonated in July 1945 and first used in war the following month, killing more than 200,000 people in Hiroshima and Nagasaki. Controversy still rages over whether the use of the atomic bomb in Japan was justified. In the more than sixty years since, it seems impossible to determine whether the deterrence that came with possession of nuclear weapons by both sides of the Cold War saved more lives by preventing a massive "hot" war or cost more lives by prolonging the Cold War. And assessing the broader social and cultural effects of "the Bomb" seems impossible. Today, the number of countries with nuclear weapons is continuing to increase, as are the risks that those weapons will fall into the hands of terrorists or other non-government actors. Each hour holds the risk that Hiroshima and Nagasaki will be displaced as the most recent use of nuclear weapons against people. Although the Manhattan Project led to beneficial medical uses of radiation and also to the controversial development of nuclear power, on balance the verdict on nuclear weapons should probably be the same as that Chinese premier Zhou Enlai is said to have given when asked about the consequences of the French Revolution, which had taken place more than 150 years earlier: "It is too soon to say."

Physicists were intimately involved not only in creating nuclear weapons but in advocating for them, as in the August 1939 letter from Albert Einstein and Leo Szilard to President Franklin D. Roosevelt that sparked the federal government's interest in such weapons. Some of the scientists involved in the development of nuclear weapons were unapologetic about the negative effects of their work, others were deeply troubled, and many were ambivalent.

What the physicists did was, at least, "good science," however one judges the results. By contrast, I believe that in eugenics geneticists were responsible for a moral fiasco based on bad science. The eugenics movement was started by Francis Galton, Charles Darwin's cousin, who believed that human evolution was beginning to move backward. In Victorian England, he saw "bad" parents having too many children and "good" parents too few. Galton's answer has come to be called "positive eugenics," encouraging "good" parents to have more children. In some places, this quickly changed into "negative eugenics," preventing "bad" parents from having children. In 1907 Indiana became the first American state to pass legislation requiring sterilization of the allegedly inferior parents; over the next twenty years, thirty more states followed it. Before eugenics disappeared in America after World War II, about 60,000 men and women were surgically sterilized by court order, for conditions such as feeblemindedness, alcoholism, insanity, epilepsy, and criminality, which have little or no genetic basis.

Gregor Mendel's groundbreaking genetics work with pea plants, published in 1866 and ignored, was rediscovered in 1900. It was then quickly and easily applied to humans—too quickly and too easily. American geneticists, led by Charles Davenport at what became the Cold Spring Harbor Laboratory, decided that the simple Mendelian framework of inheritance was important in a wide range of medical conditions, as well as traits such as "pauperism," "nomadism," "shiftlessness," and "thalassophilia" (the love of the sea). This was bad science, extending new and valuable ideas into areas powerfully affecting human lives without rigorous examination or proof.

Scientific opinion began to turn against this grossly oversimplified human genetics during the 1930s. After World War II and the revelation of the depths of Nazi use of eugenics, the movement withered away—too late, of course, for its victims. The eugenics movement extended well beyond scientists, with substantial support on both ends of the political spectrum. But the prestige and authority of science were indispensable to its influence.

Neuroscience arguably had its own ethical disaster in the widespread adoption and use of the surgical procedure usually called prefrontal lobotomy. In 1936, Portuguese neurologist Egas Moniz developed a procedure he called the prefrontal leucotomy, in which the connections between the frontal lobe and the rest of the brain were severed. Moniz initially used it mainly to treat patients with depression or other affective disorders. But its application was expanded rapidly and recklessly, particularly by Walter Freeman in the United States, where it is estimated that he and his followers lobotomized more than 40,000 patients in the 1940s and 1950s.

Moniz won the Nobel Prize in Medicine for his invention in 1949, but by the late 1950s the lobotomy had fallen out of favor. More attention was paid to its side effects, which often included major damage to the personalities and intellectual powers of patients, and alternative treatments for depression and schizophrenia became available. Although some speak positively of Moniz's work in the 1930s and 1940s, at least for carefully selected patients, the widespread use of lobotomy in the United States in the 1940s and 1950s seems indefensible.

Risks in Contemporary Brain Science

The development of nuclear weapons involved the ethically ambiguous use of good science. Eugenics was the unethical use of bad science. And the lobotomy can be seen as the extension of pretty good science past its medically, and thus ethically, appropriate limits. All three types of problems could affect the applications of neuroscience to society.

Human biological enhancement is a controversial issue on many levels, from cosmetic surgery to professional sports to selection for genetic traits. Because of the brain's importance, its enhancement raises particular concern. For millennia, humans have used psychoactive substances, from alcohol to nicotine to caffeine, to enhance mental properties, but brain enhancement is moving into new territory. Consider just two examples, one of bad science and one of good science extended too far.

Today consumers are buying ginkgo biloba, choline, acetyl L carnitine, and other nutritional supplements to improve their mental functioning. These compounds have no benefits proven by rigorous clinical trials; at best, their claims are based on limited or bad science. Yet some risks have been proven, including bleeding complications from ginkgo biloba for some people and unsafe drops in blood pressure from choline. As a result, consumers are taking unknown risks for scientifically baseless but well-advertised benefits.

On the other hand, healthy and ambitious high school and college students increasingly are using prescription drugs, not to relax or get high but to help them study. The prescription stimulants Adderall and Ritalin have important and well-documented uses in treating people with attention problems, disorders serious enough to justify the risks of these drugs. But their non-prescribed use by students in an effort to improve normal functioning applies good science past its limits, into an area where the benefits of the drugs do not justify their risks. The promised new wave of drugs to treat neurological diseases, such as the many drugs in clinical trials for Alzheimer's disease and other memory disorders, will raise a new set of issues about appropriate applications of good science.

Perhaps the most promising, and the most unsettling, area of neuroscience comes from the explosion of neuroimaging tools, which let us watch the workings of living human brains. If, as neuroscientists have come to believe (they have convinced me), the mind is produced by the functioning of the brain, close enough examination of a person's brain may allow us to know some aspects of that person's mental state—in a rough sense, to read his mind. This mind reading seems unlikely ever to reach the level of detailed thoughts, but it may already be adequate for deciding whether someone is feeling pain or anger and why. Ultimately, it may be able to say much more, including whether someone is lying.

Already, two commercial firms, Cephos Corporation and the bluntly named No Lie MRI, have announced that in 2006 they plan to offer lie detection services based on magnetic resonance imaging (MRI). Each company has licensed technology based on peer-reviewed publications by respected neuroscientists. The few published research reports, however, are based on small studies, whose subjects are usually college students presented with artificial problems. I think few outside observers believe that the effectiveness of these techniques in the real world—with real people, telling real lies—is close to established. But neuroscience-based lie detection is entirely unregulated. Soon people's lives may change, for better or for worse, because of a scientifically questionable but financially rewarding interpretation of a brain scan.

The problems caused by inadequate science, though real, pale compared with the challenges that would be posed if lie detection were scientifically valid. Effective lie detection would be the negation of what has been an effective, if not celebrated, right—the right not to tell the truth. The possible invasion of that intimate method of privacy raises a host of hard questions. Should it be allowed at all? If so, for what purposes and by whom? By the intelligence community to search for terrorists? By the military for field intelligence? By the police to look for burglars? By the government to identify illegal immigrants? By teachers who want to know if the dog really did eat the homework? And for each allowed use and user, should procedural protections be required, such as truly voluntary consent or a judicial warrant? Good science could well be used here for bad ends.

Like "mind reading," "mind control" is an inflammatory term, but our power to do just that is keeping pace with our progress in seeing into the brain at work. Some reasons to develop mind control lie in the laudable goal of helping the mentally ill, whether by lifting crippling depression or preventing psychotic hallucinations. Our increased understanding of the brain will lead to an increasing ability to "adjust" it, through pharmaceuticals or devices, thereby allowing psychologically afflicted people to lead more normal lives. The dangers of implementing such methods too soon, on insufficient evidence of safety and efficacy, always exist, as do the dangers (exemplified by the history of the lobotomy) of implementing them too broadly. And safe and effective applications may be, in some circumstances, the most frightening.

Another set of questions concerns what we should allow people to do voluntarily to their own personalities—and indeed when and whether such decisions are voluntary. But coercion raises even harder issues. The Supreme Court has already allowed mentally ill prisoners to be forcibly medicated so that they can be sane enough to be executed. Should drunk drivers be stripped of the ability to enjoy alcohol, and hence the temptation to abuse it? To what extent should people be "cured" of what they consider to be their personality traits? Should parents be able to use neuroscience to "adjust" their children, something some critics think is already happening with prescription drugs but that new techniques might make more powerful? As a parent of two teenagers, I can imagine the attraction of pills to "help them" clean their rooms or do their homework before the last minute. On the other hand, should the state be allowed to interfere in how parents choose to raise their children? What of the government, near or far, that might use neuroscience to make dissent disappear—not through the bread, sex, and soma of *Brave New World* or the propaganda and torture of *1984*, but with a little blue pill? These are not new issues or new fears, nor do they have clear answers, but the rush of progress in neuroscience gives new importance to finding workable and ethical answers to them.

These are just a few of the many ways neuroscience will raise hard questions for our society. Whole books have been written about neuroethics and its dilemmas.[2, 3] Some of the examples I have discussed may not come to pass; I suspect they will be outnumbered by the problems that no one has yet imagined.

Primum Non Nocere

I do not argue that neuroscience research should be stopped, or slowed, even in areas that might lead to abuses. To hold back on such promising research would raise its own set of moral problems. Neither do I argue that researchers must bear full responsibility for the consequences of their work, any more than parents can be charged with the full moral burden of the acts of their children. Predicting the future is easy and, if done humbly, useful; predicting the future accurately is impossible. The engineers who created the de Havilland Comet, the first jet airliner, could not have anticipated and cannot be held responsible for September 11.

But researchers can be asked to think about the implications of their work and to take reasonable steps to prevent negative consequences, for individual research subjects and for society as a whole. We cannot make *primum non nocere*, "first do no harm," a binding obligation; too often, in spite of the best motives and the most expert execution, harm will occur. But for researchers as well as for physicians, doing no harm can be an aspiration. And that aspiration can encourage us all to think about the ethical, social, and legal consequences of our work.

In February 1975, most of the leading researchers working on recombinant DNA technology, the basic method of genetic engineering, met at the Asilomar Conference Center in California and declared a moratorium on their own research until questions about its safety were answered. This meeting is a powerful example, though also a controversial one. Some argue that the conference did too much and held back important research; others insist that it did too little. But no one can deny that molecular biologists, having learned from the experience of nuclear physicists, did face directly at least some of the possible risks of their work.

The potential benefits from neuroscience are breathtaking, but so are some of the potential harms. The increasing talking, writing, and—most important—thinking being done about neuroethics issues is encouraging. This kind of work runs the risk of getting too far ahead of the science and fruitlessly piling speculation on top of conjecture. But if done carefully and modestly, with a solid grounding in the science and an appreciation of both the scientific and the social uncertainties, thinking about neuroethics may help us maximize the benefits and minimize the harms of the revolution in brain science. All of us who are interested and involved in neuroscience should feel a duty to try to accomplish those ends. I believe it is by no means too early for such a commitment. I am optimistic that it is not too late.

CHAPTER 7

The Intuitive Magician

Why Belief in the Supernatural Persists

Bruce Hood, Ph.D.

Bruce Hood, Ph.D., is the chair of developmental psychology and professor at the Bristol Cognitive Development Centre, University of Bristol, United Kingdom. His research interests include developmental cognitive neuroscience, face and gaze processing, inhibitory control of thoughts and actions, spatial representation, and magical reasoning. He can be reached at Bruce.Hood@bris.ac.uk.

HOW OFTEN HAVE YOU THOUGHT about someone you haven't heard from in a long time, only to soon receive a telephone call from that person? You may find it difficult not to think that some connection exists between the events; surely they were not simply a coincidence. Would putting on secondhand clothing that had belonged to a murderer make you feel uncomfortable? Even though the clothes have no physical trace of the previous owner, you may feel that they are contaminated. Have you ever felt that you were being watched from behind and then discovered that, indeed, someone was staring at you?

Most of us have had similar unusual experiences and sensations that defy obvious explanation. Even in this modern scientific age, many who consider themselves rational are still sometimes surprised by irrational thoughts. We may all recognize the fantastical nature of ghosts, fairies, and wizards, but other, equally magical beliefs are so common that the majority of adults assume that supernatural occurrences—those that cannot be explained by natural laws—are real. For example, a 1990 Gallup poll found that only 7 percent of Americans did not believe in any form of supernatural experience, such as telepathy, déjà vu, reincarnation, or ghosts.

But not one shred of reliable scientific evidence can be found for such phenomena. In 1979, a panel commissioned by the U.S. National Research Council to investigate psychic events concluded that "despite a 130-year record of scientific research on such matters, our committee could find no scientific justification for the existence of phenomena such as extrasensory perception, mental telepathy, or 'mind over matter' exercises . . . . Evaluation of a large body of the best available evidence simply does not support the contention that these phenomena exist." Somehow, the message of the National Research Council has fallen on deaf ears.

How can science make sense of such mass delusion? I propose that a belief in supernatural forces originates with the same mental and physiological processes that also lead us to rational explanations through what is called intuitive reasoning. By "intuitive," I mean the spontaneous, unlearned reasoning that governs many of our decisions and that often operates in the background of our conscious awareness.

The Development of Intuitive Reasoning

Intuitive reasoning emerges early in development and operates in several areas of knowledge. For the past twenty years or so, psychologists interested in the developing mind have considered intuitive reasoning to be a group of specific problem-solving mechanisms designed to manage different types of knowledge, rather than a general, all-purpose mental process. Hundreds of experiments observing very young babies indicate that humans reason about physical, biological, and psychological aspects of the world before their first birthday. How do scientists form conclusions about a baby's reasoning when language has not yet developed? Quite simply, they show the baby magic tricks.

Magic tricks fascinate us because they violate our knowledge of the world. If I hide a solid object in my hand and then open my hand to reveal that the object is no longer there, an observer expresses surprise. We all understand that solid objects do not suddenly go out of existence, even if we cannot see them. Our minds represent the object—we remember it—and so its apparent disappearance violates our expectation that it should still be present.

Using the same logic, scientists studying infants have contrived sequences of events to test intuitive reasoning. They make objects disappear or suddenly behave as if they are alive. Babies may not gasp and applaud at such tricks, but they do look longer at apparently impossible sequences than they do at similar sequences that do not break any rules about how the world works. By comparing the amount of time that infants spend looking at unexpected outcomes versus expected ones, scientists infer that infants respond differently to the two outcomes. The babies' brains must be processing and representing fundamental properties that govern the way the world operates. Scientists call such properties "intuitive theories" because, like formal theories, they provide an explanatory framework that enables the child to make sense of events as well as to predict future events.

In most cases, intuitive theories capture everyday knowledge, such as the nature and properties of objects, what makes something alive, or the understanding that people's minds motivate their actions. But because intuitive theories are based on unobservable properties (many of the properties of objects, such as the force that makes something alive or possession of a mind that controls behavior, are not visible), such theories leave wide open the possibility of misconceptions. I believe that these misconceptions of naïve intuitive theories provide the basis of many later adult magical beliefs about the paranormal.

To understand how this could happen, we need to look at the function of the brain. My argument that belief in the supernatural derives from normal brain processes includes four important points:

- The brain is designed to process and fill in missing information from a complex world of input.

- Understanding and predicting the world requires generating intuitive theories that infer invisible mechanisms to explain our experiences.

- Normal intuitive theories can bias rational individuals toward irrational reasoning.

- Not only are supernatural beliefs unavoidable (since they are anchored to normal intuitive theories), but these beliefs may confer beneficial effects by giving us a sense of control and purpose, perhaps even enhancing our creativity.

Perception: Filling in the Blanks

Our brains are designed to make sense of a world that bombards them with input. Often this information is messy, incomplete, or ambiguous, and so to organize it into meaningful patterns or perceptions, the brain makes assumptions about what the input should be and tries to fit the information into existing models. This model building is highly sophisticated in that it can make assumptions about the true nature of input even when that input is incomplete.

For example, the white square floating above four black circles that most of us see in the shapes in Figure 1 (next page), is not really there, but

Figure 1

we assume that it is, since it is the best explanation for the arrangement of elements in the diagram. Spookily, neurons in the visual-processing areas of the brain that correspond to the location where the edge of the square should be if it existed, known as "end stop cells," respond to our hallucination of the invisible edges, actually firing as if the white square is really there. Sensing an overall organization in what it perceives, the brain literally fills in the missing information.

Human perception is full of many such examples. As my colleague Richard Gregory, D.Sc., at the University of Bristol, has championed, the brain generates models or hypotheses that look for the best approximation of what exists in the real world. Even young babies do this. If you show them the pattern above, they see the ghostly square. If, after they look at this pattern for a long time, you show them a real square or some other shape, they look longer at the new shape. Neuropsychologists interpret this response to mean that the babies saw the illusory square, eventually got bored with looking at it, and so found the different shape more interesting.

The argument for a brain that not only organizes but generates input is compelling when we look further at the way the brain fills in missing information to make sense of data. For example, as you read this text you are unaware that large portions of your visual field are missing. The brain supplies these to create the illusion of completeness. None of us can see the whopping big holes in our visual field that correspond to the blind spot on each retina (it's about the size of an apple held at arm's length). Close your left eye and focus on the cross in Figure 2 with your right eye, move the page from side to side—or move your head if you

Figure 2

are looking at a computer screen—and the circle will mysteriously disappear.

You have just found the blind spot in your right eye that corresponds to the point on your retina where you have no photoreceptors (the cells that are triggered by light photons to begin the process of vision). The blind spot might be considered a design defect in the wiring of the sensory system, but under normal conditions we are not consciously aware of it. We simply cannot see what we are missing. But what I find even more interesting than this lack of awareness is the way that, as cognitive neuroscientists have shown, the brain appears to generate input to fill in the gaping hole in our visual field rather than simply ignoring it.

The brain also tackles more-complex tasks, doing much more than processing and organizing sensory input. The human brain can combine that input with memory and learning to generate explanations and predictions about the world. In other words, perception, learning, and memory together form the basis of cognition, the ability to reason.

Building Models of the World

The second point in my argument states that the brain builds theories of the world to explain many of the world's unobservable properties. For example, the world is full of objects that are controlled by invisible, external forces. When we watch two billiard balls collide, we see one forcing the other to move, but, as the Scottish philosopher David Hume pointed out in the eighteenth century, all we really see is a sequence of events, because the transfer of energy is invisible. Nevertheless, we perceive this causal force directly, and, in fact, it is extraordinarily difficult not to think of such events as involving forces. As the philosophers would say, humans are causal determinists; we cannot help but experience the world as a continuous sequence of events and outcomes. Like other aspects of perception and cognition, this way of experiencing the world may be innate. Many infant studies have shown that babies interpret events in terms of the activity of invisible forces.

Babies, like adults, use the movement of objects to determine whether the objects are alive, or even if they have minds and intentions. Consider the following scenario from a geometric soap opera created by developmental psychologists Valerie Kuhlmeier, Ph.D., Karen Wynn, Ph.D., and Paul Bloom, Ph.D., of Yale University. A green ball rolls toward the foot of a hill and then rolls hesitantly up the slope toward the summit, apparently with some difficulty. At the halfway point, along comes a red triangle from behind and pushes the ball easily up the remainder of the hill. Adults interpret this animation sequence in terms of "thinking" shapes that have mental goals and attributes. The ball has the goal of reaching the summit. The red triangle is helpful by assisting the ball as it struggles toward the top.

Notice how rich our interpretation is. After all, it is only an animation sequence, but our minds are designed to interpret even simple action sequences as being causal and, in some instances, driven by minds. Human babies, by their first birthday, are surprised when the "nice" triangle that has been helping shapes to get up the hill suddenly changes character and becomes "nasty" by pushing other shapes down the slope. It is as if the babies have assigned the triangle a personality, demonstrating that the human mind is predisposed from an early age to apply humanistic qualities to all sorts of nonhuman entities.

This tendency to attribute mental lives to nonhumans, or "anthropomorphism," explains why we get angry with our temperamental computers and talk nicely to our unreliable cars. Psychologists and philosophers ever since Hume have suggested that this tendency leads us to believe that many aspects of natural phenomena are purposeful, caused by agents with sentient minds. It takes only a short step further to think that things that "go bump in the night" are the result of some spirit or agent at work.

Another kind of intuitive reasoning that is relevant to understanding adult supernatural beliefs is "biological essentialism." Preschool children believe that living things have something inside them, some kind of essence, that defines what they are, irrespective of outward appearances. Children can therefore understand that although members of the same category may look different on the outside they share some deeper essential property. For example, children understand that something inside a dog makes it a member of the category of "dog," which is different from that of "cat." Children can even reject the idea of outward transformation in favor of an essential identity.

Twenty years ago, the developmental psychologist Frank Keil, Ph.D., of Yale University, told young children a story about a scientist who took a raccoon, painted white stripes on its back, changed the tail, and added a bag of smelly stuff. He then showed the children a picture supposedly of the transformed animal—really a picture of a skunk—and asked whether the animal was still a raccoon or a skunk. Despite the outward appearance, children reasoned that it was a raccoon, indicating that they believed in an essential element or property in the animal that could not be changed by physical appearance. Young children know nothing about DNA and genetics, but they intuitively reason that such an element defines the true identity of living things.

Generating Supernatural Beliefs

Let us consider how such intuitive theories underlie adult supernatural beliefs. Take the concept of essence, for example. Psychologist Paul Rozin, Ph.D., of the University of Pennsylvania, showed that this concept might explain the belief in psychological contagion; the idea that a nonphysical, usually negative state that we associate with the mind can transfer to objects. Initially Rozin was studying the development of disgust. He learned that if a sense of revulsion toward certain typical categories (such as feces, dirt, illness, disease, or putrefaction) had been established in early childhood, adult participants in his study would still be unwilling to touch objects that they believed to be contaminated by a disgusting item, even though the object had been thoroughly disinfected and cleaned. They perceived that the essence of the contaminant was still present. They even avoided objects that had never been in actual contact with a contaminant, such as a brand-new bedpan, thus demonstrating that the association was too strong to be overcome by rational processes.

As a general rule, such intuitive reasoning not only increases in strength with experience but becomes generalized. It is also difficult to control consciously. Recent brain imaging work by Princeton psychologists Lasana Harris, Ph.D., and Susan Fisk, Ph.D., has revealed a neural response to disgust. For example, when they showed undergraduate students pictures of disgusting people and objects while the students lay in a functional magnetic resonance imaging scanner, they found that the images activated the amygdala and insula systems of the brain, which are associated with nausea.

Rozin's work with psychological contamination is particularly relevant to understanding supernatural thinking. Not only would participants in his research studies avoid contact with items that were associated with tangible contaminants, but they did not want to wear or even touch items that they believed had once been worn by murderers. It was as if evil, a moral stance defined by culture, had become physically manifest and infected the clothing. I recently replicated this effect at a public lecture at the Dana Centre in London. Everyone in the audience was willing to put on a secondhand cardigan I held up if I promised to pay them $30 for doing so. But all but one member of the audience declined the invitation when I told them that a notorious murderer had once worn the sweater.

A belief that a psychological state such as evil can manifest as a physical entity explains many peculiar supernatural beliefs. For example, the idea of spirits and souls appearing in this world becomes more plausible if we believe in general that the non-physical can transfer over to the physical world. It explains why some believe that sentimental objects or voodoo dolls can be invested with supernatural powers.

Another good example of an intuitive theory underpinning adult supernatural beliefs can be found in examining the common assumption that we can detect someone staring at us from behind even though we cannot see them. In 1898 the Cornell University psychologist Edward Titchener, Ph.D., reported that 90 percent of his students believed they could detect the unseen gaze of others. The belief is still so common that most people are unaware that it is controversial. In a recent sample of my undergraduate class, I found that the same proportion (90 percent) of the students believed they could detect an unseen gaze. I think this belief can be explained by a combination of the natural human supersensitivity to gaze, children's intuitive theory of how vision works, our subjective experience of shifting our gaze, cultural endorsement of the belief, and, finally, poor evaluation of the evidence.

In primates, including humans, gaze is the primary channel of nonverbal communication, with neural structures dedicated to faces in the brain's visual processing areas. In addition, gaze is extremely arousing for adults and activates the amgydala in the brain's emotional center. That is why strangers do not look each other in the eye in confined spaces such as an elevator. We talk about "exchanging glances" and a "piercing gaze" as if a physical force leaves the eyes, and many cultures draw attention to the supernatural power of the eyes, as in the "evil eye" beliefs of some Mediterranean cultures. The most common intuitive theory of vision among children is that an energy beam leaves the eyes, a belief that is perpetuated in our culture by comic heroes and cartoons. Even the subjective experience of gaze gives the impression that, as we move our eyes to scrutinize the world, we are the origin of the changing visual scene and so gaze must exit from the eyes. This "extramission" theory of vision is actually as old as Plato and was the dominant scientific theory of vision until the great Arabic scientist Al-Hazin (965–1029 C.E.) demonstrated that vision worked by light entering the eye. With education, most adults come to understand that vision actually works by energy entering the eyes, yet the belief that we can sense someone's hidden stare increases from childhood onward.

Intuitive theories are difficult to eradicate, because we are often not conscious that they are working away in the background. If you combine the intuitive sense of being stared at with the multiple times you recall your hunch was proven to be true, then this belief is naturally going to gain credibility in your mind. We do not take note of all the times when we have a sense of being looked at but, as it turns out, are wrong. Titchener, in his original paper on unseen gaze, provided another clever explanation: If we feel we are being stared at and we turn around to see who is looking at us, anyone behind us is likely to return our stare—which leads us to the false conclusion that we actually did detect their unseen gaze.

Sensing Connections and Patterns in the World

When we consider the way the brain fills in missing information, generates causal explanations for events, and builds intuitive models of the invisible properties of the world, the origins of supernatural thinking start to become apparent. Any example of supernatural belief can be viewed as a misinterpretation of the available evidence or as the assumption of patterns, forces, or essences when in fact none exist. In other words, those who infer supernatural activity are detecting order and structure in the incoming information when there may be, in fact, only simple noise.

Moreover, the brain cannot handle random patterns. If I were to take a handful of coffee beans and scatter them across the tabletop, the brain of anyone observing the scene would spontaneously and effortlessly organize the spilled beans into patterns. Likewise, because of our stubborn tendency to see outcomes as being connected to the events that precede them, we have a hard time accepting that sequential events are not causally related. For some people the sense that events are both structured and connected is quite distorted. In 1958, the German psychiatrist Klaus Conrad coined the term "apophenia" for an "unmotivated seeing of connections" accompanied by a "specific experience of an abnormal meaningfulness." Apophenia is a well-recognized symptom of schizophrenia. When people with schizophrenia experience florid hallucinations, they have a tendency to interpret random events as not only meaningfully connected but often related to themselves, a tendency that forms the basis for paranoid delusions.

This propensity to detect co-occurrences and patterns lies at the heart of many supernatural assumptions. The neuropsychologist Peter Brugger, Ph.D., from the University of Zurich, recently proposed that it stems from the relative activity of the neurotransmitter dopamine in the left and right cortical hemispheres of the brain. Brugger and his colleagues have consistently found that excessive dopamine activity in the right hemisphere is associated with apophenia and the assumption of supernatural forces.

To test his hypothesis, Brugger identified two groups of individuals—those who tended to believe in supernatural phenomena and those who were skeptics. To measure the study participants' sensitivity to patterns, Brugger administered a perception test using images of real faces, scrambled faces, real words, and non-words. Real faces and words have an identifiable pattern, but scrambled faces and non-words do not. Study participants had to pick out the real words and faces as a series of images was flashed on a screen, too quickly for them to look at them carefully. As Brugger expected, the skeptics scored lower than the "believers"; they were more likely, for example, to call a real face a non-face. But after he administered L-dopa, a precursor to dopamine, to both groups, he found that the skeptics were significantly more likely to identify faces and words that they would have previously rejected as non-patterns.

Brugger believes that the activity of dopamine is related to the brain's ability to discriminate between meaningful signals and noise and that those individuals in whom this neurotransmitter system is overactive are more inclined to see causal connections and patterns. Though all people experience this inclination to some extent, it is elevated in psychiatric disorders such as schizophrenia and in temporarily distorted brain states resulting from the abuse of psychoactive drugs that affect the dopaminergic system, such as Ecstasy.

Why Supernatural Reasoning May Be Beneficial

Rationalists such as the philosopher Daniel Dennett, Ph.D., of Tufts University, and the biologist Richard Dawkins, Ph.D., of Oxford University, have recently lamented what they see as an increase in supernatural thinking in modern society and have accused various religions of propagating myths and fairy tales in the minds of gullible young people. They are particularly concerned about the rise of religious fundamentalism and the creationist movement. I would argue, however, that most supernatural beliefs are an end product of normal intuitive reasoning in children, and so it is not clear how to eradicate such a natural mode of thought.

By advocating the need to abandon irrationality in favor of rationality, Dennett and Dawkins also may have undervalued the advantage of a mind that infers the existence of patterns, forces, and essences that do not really exist. As we have seen, such natural intuitive reasoning supports both rational and irrational models of the world. Brugger has pointed out that apophenia and creativity may be two sides of the same coin; the tendency to sense connections where other people do not is a potentially creative aspect of magical thinking.

Supernatural beliefs also give us a perception of control in situations where in fact we may have none. In his seminal paper of 1948, "Superstition in the Pigeon," the Harvard University experimental behaviorist B. F. Skinner, Ph.D., reported that pigeons rewarded with food pellets on a random basis soon settled into repeating movements that they associated with the outcome. Humans also repeat behaviors that they believe may affect positive and negative outcomes, even when there is no actual relationship between the behavior and its result. Every superstitious sportsman knows how important it is to his performance to indulge in a special ritual or wear a lucky talisman before a crucial match. This behavior is self-reinforcing, because people who are thwarted in performing the superstitious activities become more stressed, thereby affecting their ability to perform optimally. In early psychophysiological studies of the effects of stress in the 1970s, researchers learned that both animals and humans needed the perception of control, even if illusory, to make stressful events less stressful. This may explain the reported increase in superstitious rituals among residents of the most at-risk neighborhoods of Tel Aviv, compared to those in low-risk areas, during the Iraqi SCUD missile attacks of the 1991 Gulf War.

Most important, a belief in the supernatural can give people a deep sense of connection with the past and with each other. Such beliefs impart a consideration of the possibility that the mind will outlive the body. They are common to a variety of religions, but even atheists can benefit from a sense of the supernatural if belief in a deeper reality to existence shields them from facing the existentialist crisis of thinking that life has no purpose or meaning. Certainly being a scientist does not exclude one from supernatural beliefs. In her 2006 survey of 1,600 scientists from 21 elite U.S. universities, sociologist Elaine Howard Ecklund, Ph.D., of Rice University, found that only 38 percent of natural scientists and 31 percent of social scientists did not believe in God. Religious scientists, such as the geneticist Francis S. Collins in his recent book *The Language of God,* have attempted to make logical arguments for the existence of God, but such arguments are unlikely to satisfy those who demand objective evidence as proof.

Before science, the dominant explanatory model of the world was what is called "natural magic," the idea that everything was controlled by hidden mechanisms that God had put into the universe. The wise men, or magi, who sought to understand these occult forces later became the alchemists, dabbling in potions, who would eventually become today's experimental scientists. Indeed, the ability to imagine the unobservable has produced biological and computer science advances taking us right to the brink of "mind over matter." In July 2006, newspapers worldwide reported on a paralyzed research volunteer who has learned to control a computer cursor by thinking about it. But the "how" of it is hardly a case of staring and wishing hard enough to make the cursor move. The researchers guiding the project implanted signal-detecting wires in the subject's brain near neurons known to fire to initiate body movement, computer programmers translated the signals into instructions for the cursor, and the patient learned how to calibrate his thoughts of moving well enough to guide the cursor. But the process of science is by and large opaque to the lay public, so many adults still rely on intuitive theories to explain the uncanny events that occur in their lives and to discover a deeper sense of reality. As social animals with evolved inferential reasoning, we cannot avoid the magical beliefs that we fail to recognize because of our rational blind spots.

Finally, we must recognize that science, too, can benefit from a leap into the unknown by looking for structures and mechanisms in the universe that underlie the fabric of reality. We must remember that the scientific method is not a recipe for unraveling the structures of the universe but rather a process that requires systematic evaluation of theories combined with creativity and a bit of luck. Even scientists should try to imagine the impossible sometimes.

CHAPTER 8

Bringing the Brain of the Child with Autism Back on Track

Diane C. Chugani, Ph.D., *and* Kayt Sukel

Diane C. Chugani, Ph.D., is a professor of pediatrics and radiology at Wayne State University School of Medicine and codirector of the Positron Emission Tomography Center at Children's Hospital of Michigan, where her research focuses on neurochemical mechanisms in children with neurodevelopmental disorders, including autism. She is a member of the scientific advisory board for the Autism Society of America, Cure Autism Now, the Tuberous Sclerosis Alliance, and the Sturge-Weber Foundation. She can be reached at dchugani@pet.wayne.edu.

Kayt Sukel is a freelance writer whose essays and articles have appeared in *Science*, *Memory and Cognition*, and *Neuroimage*, as well as the *Washington Post*, the *Christian Science Monitor*, and *National Geographic Traveler*. Currently living in Hammersbach, Germany, she can be reached at ksukel@hotmail.com.

MANY PARENTS OF AUTISTIC CHILDREN speak about their child in terms of "before" and "after." They reminisce about their child's babyhood, full of play, smiles, and the early developmental milestones so easily reached. But then, they say, seemingly overnight, things dramatically changed.

The lively sounds that so closely resembled speech never quite evolved into actual words; instead they regressed into grunts and nonsense noises. The child who was once content to be held in Mama's arms now recoiled from her touch, and any small change in routine could result in inconsolable screams. Many say that it was almost as if their child had suddenly and mysteriously changed into an entirely different person.

Parents of autistic children usually witness this profound and often abrupt metamorphosis in their son's or daughter's development at some point between the one- and two-year marks. They wonder if perhaps the disorder could have been avoided had they caught it earlier or done something—anything—differently during that critical period. So far, the answer has always been no. New research, however, may change that answer.

The Potential for Treatment, Not Management

In this article we will explain why we believe that discoveries made using molecular neuroimaging offer the promise of a new approach to treating autism during critical periods of brain development. In the past two decades, scientists have made substantial advances in understanding autism and how it affects brain development and behavior. Research in genetics, functional neuroimaging, and cognitive neuroscience has provided helpful knowledge about potential causes of autism, as well as the range of behavioral effects. Although these studies have been invaluable in understanding more about the disorder, it is research at a more basic level that may provide the start for a new and unique method of treatment.

What is called molecular neuroimaging allows researchers to measure biochemical changes in the brains of living humans. One powerful technique, positron emission tomography (PET), measures the location of a radioactive tracer as it travels through the body, including the brain. By selecting the right biochemical from the more than 1,400 that have been radioactively tagged, researchers can measure such processes as the metabolism of glucose, how proteins are synthesized in different areas of the body, blood flow, and how neurotransmitters bind to neural receptors. For example, by using PET to track a radioactive tracer that is converted in the body to a specific brain chemical such as the neurotransmitter serotonin, researchers are now able to examine how much of that neurotransmitter is made in a person's brain. The ability to examine the brain of living subjects at such a fine level of detail allows greater understanding of the processes that underlie normal brain function and presents an opportunity to see how we might intervene if those processes are not functioning correctly.

Recent research using molecular neuroimaging has provided critical information about how the processes in the developing brain of an autistic child differ from those in the brain of a child without the disorder. Previously, very little was known about the biochemistry of the developing brain, but structural and molecular neuroimaging studies have provided us with significant clues by showing dramatic differences in brain growth and levels of the neurotransmitter serotonin in children with autism. Other research has also provided insights into the autistic brain by showing differences in the organization of neurons in the cortex and decreased numbers of certain types of cells that make up the cerebellum.[1,2] Careful study of the details of these differences in brain development may provide a basis for designing new drug treatments for autism, treatments that would not manage behavior, as is the case now, but rather would work directly to bring the brain development of the autistic child closer to the normal range.

Multiple Causes, Multiple Effects

One of the biggest problems in succinctly defining autism—and consequently in recommending treatment—is that the disorder exhibits such a broad array of characteristics. Autism is defined by the presence of multiple communication, social, and stereotyped behavioral difficulties that begin before three years of age. Furthermore, autistic behaviors vary according to both developmental level and chronological age. Problems in communication can range from no language at all to more-subtle language difficulties, such as delay in word acquisition or pronoun reversal. Stereotyped behaviors such as hand flapping and spinning are typical of young children with the disorder, whereas older, high-functioning people with autism often evidence a need for sameness or a focused interest in a particular activity or topic. In addition, many autistic children suffer from a host of other complaints, such as extreme reactions to sensory stimuli that would not bother most people and aggressive or self-injurious behavior.

Underlying this spectrum of behaviors are undoubtedly multiple causes, only a small fraction of which have thus far been identified. Recent research has shown that autism may be a component of several genetic disorders, including fragile X syndrome, phenylketonuria, tuberous sclerosis complex, and Rett syndrome. But even in cases where genes are identical (such as when identical twins have the disorder), the symptoms may manifest themselves in profoundly different ways—for example, with one twin showing delay in language acquisition while the other struggles with a sensory impairment.

Despite the diverse causes that may underlie autistic behavior, we cannot exclude the possibility of a few common neurochemical features. One of these may involve the synthesis of serotonin, a neurotransmitter critical to normal development in the brain.

Brain development involves a series of biochemical "programs" that are turned on and then turned off as the child's body builds itself. Just as the building of a house is accomplished through a series of steps—laying the foundation, erecting the beams, adding the bricks and mortar, and so on—the brain begins with the birth of neurons (neurogenesis), the formation of the connections between those neurons (synapses), and the refinement of electrochemical networks as children interact with the environment through their five senses. As a result, not only are children's brains different from those of adults, but the brains of infants are markedly different from those of toddlers, and those of toddlers are very different from those of older children and adolescents. Moreover, the programs can go awry at any point.

Developmental disorders, including autism, result from errors in the normal sequence and duration of the programs of brain development. These changes may result from a variety of genetic or environmental causes, but the disorder is still characterized by those deviations. Information about the deviations provides us an opportunity to seek a way to intervene in the developmental process at some critical point and help it get back on track.

The Benefits and Risks of PET

Using PET, researchers have studied the biochemical changes crucial to normal development in the brains of infants and children at various ages. In the majority of the PET studies published thus far, the subjects were children who had a variety of neurological or neurodevelopmental conditions, with comparisons being made among groups of children with different disorders. The optimal design for such studies, however, would be to compare experimental results from a group with a disorder—such as autism—with results from an age-matched group of children who do not have any neurological impairment.

PET imaging studies do carry some risks, among them possible negative effects from radiopharmaceuticals, the use of sedation, and exposure to ionizing radiation. Thus, while the best comparison group for research would be children without the disorder, some people hold that using the radioactive dye necessary for PET scanning in normal children is unethical. However, what is required for PET scanning is a single dose of the radiopharmaceutical in tracer quantities, which would be unlikely to result in any adverse pharmacological effects. In fact, you are more likely to find a greater number of toxins in a glass of tap water, as typical doses are lower than the quantities of toxins legally present in our drinking water. Even in a case where multiple scans might be necessary, PET scanning would still involve minimal risk to the subject. The average lifetime risk of dying from cancer is 23.66 percent for men and 19.99 percent for women, so the additional lifetime cancer mortality risk of 0.0005 from the radioactive exposure during a PET scan seems minimal, hardly more than that from events of daily life, such as getting sunburned at the beach.

Given the critical information that these studies can reveal about brain development and, of course, the profound differences in the brain at different ages, it would seem that the benefits of the procedure far outweigh the risks.

Serotonin in the Autistic Brain

Although evidence exists for the potential involvement of several neu-rotransmitters in autism, research has shown consistent abnormalities in-volving the neurotransmitter serotonin in brain development programs.[3] We have chosen to focus on serotonin here as one particularly promising avenue for investigation.

In 1961, Richard J. Schain, M.D., and Daniel X. Freedman, M.D., first reported increased blood serotonin levels in approximately one-third of people with autism, but the less-sophisticated technology available at the time made it difficult to demonstrate whether the change in serotonin in the blood also signified changes in serotonin in the brain. Recent-ly, however, using PET scanning, our research team (D. Chugani) has detected differences in serotonin production between autistic and non-autistic children.[4] During early childhood, children without autism under-go a period of high brain serotonin synthesis, which then declines when they are about six years old. We hypothesize that this period of higher se-rotonin synthesis helps children form strong new synaptic connections in their brains. In autistic children, however, this process is disrupted. In our studies, the autistic children did not show such an age-dependent peak and decline but instead showed a consistent rate of serotonin synthesis in the brain—underproducing serotonin during the critical early years and overproducing the neurotransmitter after the age of six.

Because of this flattened level of serotonin synthesis in autistic children, without the early critical peak that we see in children who do not have the disorder, we hypothesize that young autistic brains are not making as many, as strong, or as accurate synaptic connections between neurons.[5, 6] This could explain many of the behavioral problems observed in autistic children, including extreme reactions to sensory stimuli, for example. Some autistic children are unusually sensitive to fluorescent lightbulbs, responding hysterically to the bulbs' flickering stimuli, which might not even be noticed by a child without autism. This extreme reaction could be the result of incorrect routing of a synaptic connection during brain development, which would pair the stimulus with an uncommon reaction.

Treatment Now and to Come

Currently, most treatments focus on managing behavior through drug or behavioral therapies. For example, Risperdal, one of the newer antipsychotic medications prescribed for schizophrenia, bipolar disorder, and agitation, has been found to be effective in treating aggression and sleep problems in autistic children. On the behavioral side, a form of therapy called applied behavior analysis focuses on teaching new skills, as well as correcting undesired habits, by breaking behaviors down into small, manageable steps and increasing a desired behavior with positive reinforcement. Unfortunately, these therapies have limited success. Moreover, they simply treat the symptoms, not the cause.

Traditionally, researchers have been reluctant to consider drug therapies for young children, in order to avoid disrupting the normal series of events in their rapidly developing brains. But in the case of children with autism, that series of events does not seem to occur as it should. Therefore, a course of drug treatment during early childhood could potentially alter brain development and correct the deviation from the normal program, thus having a lasting impact on the organization of the brain and guiding its development to more closely mimic that of non-autistic children.[7, 8]

Although many drugs are used to treat various diseases in children, most of them have actually been tested only in adults, not in children. And those drugs that have been tested in children have rarely been tested in very young children—even drugs routinely used to treat those children for diseases. We do need to protect children from the risks inherent in the testing and research process, but the risk of treating children with drugs that have been tested only in adults is also substantial, because the biology of the developing child is so different from that of the adult.

As we consider using drugs to treat developmental disorders such as autism, therefore, we must keep in mind that the brains of a newborn infant, an eighteen-month-old toddler, and a four-year-old have dramatic differences in the need for energy, the concentrations of neurotransmitters, and even the patterns and locations of neurotransmitter receptors. Because of these dynamic differences, the use of a drug that affects a receptor critical in the development of a particular brain function can actually dramatically change the course of brain development. In the past, such an effect was typically viewed as toxic—as damaging the brain, sometimes permanently. More recently, however, the possibility of using drugs during critical periods of brain development might be a way to take advantage of "windows of opportunity" to put brain development back on course.[9] Molecular neuroimaging has provided us with such a unique window by helping us understand more about how the brains of autistic children develop.

A Challenge to Autism Researchers

A growing body of evidence indicates that serotonin regulates several aspects of brain development, including cell division and differentiation, the growth of neurites on neurons, the creation of new synaptic connections and possible critical periods for development, and what is called activity-dependent plasticity, or when a person's interaction with the world can help the brain develop. Given this understanding of serotonin's effect on brain development and the discovery that autistic children lack adequate amounts of this neurotransmitter during early critical periods of brain development, it would seem that we have a unique opportunity to provide a drug treatment that would help regulate serotonin levels in the autistic brain and help deviating brain development to get back on course.

Because using drugs in young children is likely to have a powerful impact on developmental processes such as synapse formation and elimination, the drugs may be particularly effective during the period of behavioral regression that typically occurs in children with autism when they are between twelve and twenty-four months of age—a very dynamic period for the creation of synaptic connections as well as the age period where autistic symptoms are first observed.[10] Young children have many, many more cortical synapses in their brains than adults do. As they get older, this number decreases; some synapses that are not being used are eliminated, while those that are being used are maintained. Although scientists do not yet know the precise mechanisms responsible for this initial bounty of synaptic connections, they do know that changes in levels of neurotransmitters during developmentally critical periods can result in altered creation and maintenance of synapses.

In addition, something as simple as a young child's interactions with the environment and other people can potentially create new and diverse synaptic connections.[11] Spurred by these activities, synapses are created, strengthened, and stabilized. One process by which synapses are believed to be stabilized is called long-term potentiation.[12] During development, serotonin affects long-term potentiation in several areas of the brain, among them the somatosensory cortex, the visual cortex, and the hippocampus, as well as in the spinal cord.[13] This is particularly interesting since these parts of the brain are involved with sensory processing, an area where many autistic children have difficulties.

Designing the What, When, and How

Using drugs that mimic the effect of serotonin in young children with autism may be beneficial in encouraging their brains to develop more normally. As we begin to design studies to develop such a treatment, however, we encounter many critical decisions: the choice of the drug, what subjects of what age to include, the duration of treatment, and whether or not the drug is paired with a corresponding behavioral intervention. Several classes of drugs specific to serotonin function could prove to be useful, among them serotonin uptake inhibitors (such as Prozac) or agonists (drugs that act like serotonin in the brain).

Although as a group young autistic children (under six years old) in whom serotonin synthesis was measured with PET showed lower serotonin synthesis than non-autistic children, the values varied. Children with the lowest values might benefit from treatment more than children whose serotonin synthesis is only slightly lower than normal. Once again, molecular neuroimaging can help us, if we use PET scanning to identify those children who might benefit from treatment with a serotonin-like drug at an early age.

In order to determine the best time to treat an autistic child, we must consider both the normal developmental process and that child's deviation from it. Some interventions may work in autistic children only during a certain age range or during a certain period of brain development. In general, one would hypothesize that the earlier the intervention and the closer in time to the actual deviation in brain development, the better the outcome will be, whether this means recovery of lost function or gaining a function that might not have ever fully developed. Since serotonin synthesis has been shown to be lower in autistic children between two and six years of age, it seems that treatment should begin as soon as a child is diagnosed with autism. The earlier the disorder is diagnosed, the more effective the intervention is likely to be. This is something of a balancing act, however, since the side effects of drug treatment may be less significant at a later time.

The question of how long to treat an autistic child is similar to that of when to treat. Once again we need insight from molecular neuroimaging into the normal course of development and the deviation from it in autism. But treatment for longer than the specific period during which a particular brain process is disturbed could be beneficial. For example, it might lengthen the critical period, providing more time for sensory interventions to affect the strength and frequency of synaptic connections.

If this is the case, the length of treatment might be determined by observing when the side effects begin to surpass the perceived benefits.

Because interaction with the outside world can strengthen synaptic connections, we should also consider combining behavioral interventions with drug treatment. Since serotonin-facilitating drugs alter the formation of synapses in the somatosensory cortex, pairing them with behavioral therapies that target sensory impairments may improve the efficacy of the behavioral intervention. Conceivably, drugs could also be used to lengthen or reopen critical periods in brain development, allowing more time for acquisition of certain skills, such as those involved in language. Pairing the opening of the critical period with intensive behavioral therapies could lead to the possibility of the synaptic connections' being established outside of the normal period. For example, in an autistic child with language problems, pairing a behavioral language treatment with a drug that facilitates serotonin transmission may create the right environment to encourage the growth of stronger, more accurate synaptic connections and, therefore, help the child improve speech capability and comprehension.

Drug Intervention: A Paradigm Shift

Using drugs to treat children, especially very young children, is always a matter of concern. But the greater understanding of brain development that has been gained through molecular neuroimaging suggests that pharmacological treatments may allow us to guide the brain development of autistic children to match more closely that of their normal peers. This approach could help rein in developmental deviations in children with autism, giving them the ability to recover lost skills as well as to gain new ones more quickly. For the parents of these children, who have had to helplessly watch them succumb to the symptoms of autism, this simple paradigm shift could provide a way to treat the disorder before the autistic metamorphosis is complete and the familiar baby they knew is lost to them forever.

The benefits of such a new, focused treatment for autism far outweigh the perceived risks. New drug treatments being developed that facilitate the function of serotonin, paired with behavioral therapies, hold the promise of a radically different developmental outcome for children with autism and, down the road, for children with other developmental disorders as well.

CHAPTER 9

Toward a New Treatment for Traumatic Memories

Jacek Dębiec, M.D., Ph.D., *and*
Margaret Altemus, M.D.

Jacek Dębiec, M.D., Ph.D., is a psychiatrist, neuroscientist, and philosopher. He is a research scientist at the New York University Center for Neural Science, where he studies the neural mechanisms of emotional learning and memory. He is the coauthor with Joseph LeDoux and Henry Moss of *The Self: From Soul to Brain* (New York Academy of Sciences, 2003). He can be reached at jd86@nyu.edu.

Margaret Altemus, M.D., is an associate professor of psychiatry at Weill Medical College and a board-certified psychiatrist. Her research focuses on the role of the endocrine system in affective disorders and the physiology of the stress response in humans. She can be reached at maltemus@med.cornwell.edu.

THE TITLES OF STORIES for non-scientists about research on altering traumatic memories express the hopes and fears of our society:

- "Studies say old memories can be lost" (Carey Goldberg, *Boston Globe*, 2003)
- "Blank for the memories: Someday you may be able to take a pill to forget painful recollections" (Scott LaFee, *San Diego Tribune*, 2004)
- "Is every memory worth keeping? Controversy over pills to reduce mental trauma" (Rob Stein, *Washington Post*, 2004)
- "When remembering might mean forgetting" (Douglas Steinberg, *The Scientist*, 2004)
- "Rewriting your past: Drugs that rid people of terrifying memories could be a lifeline for many. But could they have a sinister side too?" (Gaia Vince, *New Scientist*, 2005).

Some of these stories' authors, or at least the headline writers, have stretched the current science a bit. Forgetting, for example, is an active psychological process, not a simple memory erasure; and none of the studies so far has demonstrated a complete blockade of a targeted memory. But these writers are raising some of the right questions.

Scientists have made great progress in understanding the neural basis of learning and memory, and their discoveries suggest it might be possible to use drugs to relieve the distress of traumatic memories. For those who carry memories too painful to bear, such as people who suffer the nightmares and intense flashbacks that characterize post-traumatic stress disorder (PTSD), a method to block oppressive recollections is well worth the effort. The National Center for PTSD estimates that, over a lifetime, 7.8 percent of adult Americans will suffer from intrusive, often disabling memories associated with PTSD. Those with PTSD often relentlessly avoid anything that might trigger memories of a trauma. Why not help them?

But as the headlines suggest, there are ethical implications to altering one's consciousness through drugs. Which memories should be clouded? Whose? How much? Such public attention also has spurred debate among scientists, philosophers, ethicists, lawyers, and lawmakers about how and when one might use drugs to alter memories. As practitioners and as a society, we need to ask, What should the limits be?

How Memories Form

All memories are made in stages. The initial phase of new learning is often known as short-term memory. In this stage, newly acquired information is unstable and susceptible to interference, such as when a brief distraction makes us instantly forget the phone number we have just learned. If undisturbed, new learning becomes consolidated within a few hours into a long-term memory that is much more resistant to interference.

Research in animals has shown that the development of long-term memories involves activation of molecules at the synapses between nerves. This initiates cascades of intracellular reactions that modify the synaptic connections. Drugs that disrupt any aspect of consolidation prevent stable, enduring memories from forming, yet once the window of consolidation is closed, the same drugs do not have any effect on the memory. Similarly, drugs that enhance memory consolidation do not have the same result if they are administered more than a few hours after the new experience.

Learning and memory also are influenced by many natural factors, including stress and emotional arousal, both of which involve a release of norepinephrine (also known as noradrenaline). The main source of norepinephrine in the brain is the locus coeruleus (in Latin, "blue spot"). The locus coeruleus, located in the brain stem, sends abundant projections to other parts of the brain, including the amygdala, the hippocampus, and the prefrontal cortex, all of which play important roles in memory formation. From animal studies, we know that stimulating one type of norepinephrine receptor at nerve synapses—the beta-adrenergic receptor—enhances the intracellular processes that contribute to memory consolidation and thus strengthens memories, and that blocking it during stress or arousal prevents its augmenting effects on memory.

Norepinephrine acts in several brain sites to strengthen memory formation. The release of more brain norepinephrine or more intense activation of the locus coeruleus inhibits performance of the prefrontal cortex, which plays a role in emotional control and extinction or suppression of memories. This stimulation also excites the amygdala, the key part of the brain in generating fear behaviors. The combined effects of norepinephrine on the prefrontal cortex and the amygdala may explain why we sometimes acquire habitual reactions that are difficult to control.

A trauma, by definition, is associated with high levels of arousal and activation of stress hormones. Many scientists interpret the clinical symptoms of PTSD, such as nightmares, flashbacks, and increased arousal in response to trauma-related cues, as an exaggeration or disturbance of the normal processes of emotional learning and memory. In this view, PTSD may be understood as a consequence of overconsolidation of the traumatic memory caused by increased activity of stress hormones and other biochemicals. In a 2002 clinical study, Roger Pitman, M.D., and his colleagues from Harvard University first applied this idea to people who had experienced a trauma.[1] They administered propranolol (a beta-adrenergic receptor antagonist commonly used for treating hypertension and cardiac arrhythmias) to people who came to an emergency room after experiencing a trauma, and they continued the drug treatment for ten days. When they tested these study participants three months later, they found that the propranolol had lowered the risk of developing hyperarousal in response to cues that reminded participants of their particular trauma. This approach has limits, though: the risk of developing PTSD after most types of trauma is relatively low, and many people who have experienced a trauma do not immediately seek medical treatment.

To create treatment strategies that can be used long after PTSD has developed, researchers are now turning to what has been learned about the biological underpinnings of processes known as memory reconsolidation and extinction.

How Memories Reconsolidate

Until very recently, most brain scientists believed that memories were completely immune to pharmacological alterations once they were consolidated. But there were a few animal studies in the late 1960s and early

1970s that suggested otherwise, reporting that administering a consolidation-blocking drug shortly after a reminder of a long-term memory impaired the subsequent strength of that memory. In 1979, Donald J. Lewis, Ph.D., from the University of Southern California, proposed a distinction between active and inactive states of memory, based on the earlier reports.[2] Perhaps retrieving well-established memories activates them in a way that renders these memories vulnerable to either disruption or strengthening by drugs that act on the underlying neurobiological systems.

Two decades later, Susan Sara, Ph.D., from Centre National de la Recherche Scientifique in Paris, built on Lewis's ideas. To explain how drugs might cause deficits in a previously well-established memory, she suggested that reactivating a memory by recalling it triggers another round of consolidation, which she labeled reconsolidation. This reconsolidation might enable the memory to be updated, incorporating new experience.

To test this notion, Karim Nader, Ph.D., Glenn Schafe, Ph.D., and Joseph E. LeDoux, Ph.D., from New York University, investigated fear memories in rats, using a common experimental learning model called auditory fear conditioning and what they knew about where in the brain such memories are formed. In this model, animals hear a tone—the conditioned stimulus—then feel a mild electric foot shock—the unconditioned stimulus. Animals learn that the tone precedes the shock, and when researchers subsequently play the tone alone, the animals freeze out of fear. By measuring how long the animals freeze, researchers estimate the strength of their fear memory. In auditory fear learning, the amygdala is a key brain structure. Previous studies had shown that infusing the amygdala with consolidation blockers, including the protein synthesis inhibitor anisomycin, disrupts formation of long-term auditory fear memories.

Using this model, Nader and his colleagues trained their rats to be scared of a tone. After enough time had passed that the fear memories had consolidated, they sounded the tone again without the shock and immediately infused anisomycin straight into the rats' amygdalas. The next time they heard the tone, rats that received the anisomycin froze for a much shorter time, expressing less fear. Interestingly, the drug caused this apparent amnesia only if it was infused immediately after the tone; it had no effect when administered a few hours later or when not preceded by the tone that reactivated the rat's memory.

Since 2000, when Nader and his colleagues published their findings in the journal *Nature,* other scientists have published studies showing that reconsolidation occurs in a variety of species, including the snail, sea slug, crab, honeybee, and mouse, as well as the rat. Recent studies also suggest that although reconsolidation and consolidation share similar mechanisms, they are distinct molecular processes. Both may be altered, though, for better or worse, by administering a drug.

In the first human-based study, published in 2003, Matthew Walker, Ph.D., and his colleagues from Harvard University trained participants in the study to perform a short sequence of finger movements. Once the learning had consolidated, they asked participants to repeat the sequence and then immediately instructed them to perform a different manual task (called the interference task). Walker and his colleagues observed that interfering with the reactivated memory of the finger sequence profoundly impaired people's future performance of that sequence. No drugs were used; instead memory was altered by a behavior that interfered with the same neurobiological processes.

From Animals to Humans

Scientists are eager to apply these discoveries to develop new treatments for PTSD and other forms of disabling fear, but the studies so far are not a one-to-one match. The demonstration by Walker and his colleagues that a consolidated, long-term human memory can be disrupted is exciting, for example, but the finger-sequencing task may not involve the memory systems that are crucial to traumatic memory. A difference may well exist between how conscious declarative memories are formed and re-formed and procedural forms of learning, such as the finger task. In addition, the finger-sequencing task does not result in an emotional memory that would be associated with releasing stress hormones, let alone with traumatic, intrusive features.

Researchers are now focusing on translating to humans the animal studies that used drugs to lessen the intensity of fear memories. Some re-consolidation blockers, such as protein synthesis inhibitors, are very tox-ic, but other drugs can be safely used in humans. In 2004, Jacek Dębiec and Joseph E. LeDoux showed that propranolol can block reactivated conditioned fear responses in rats.[3] It was effective in disrupting recon-solidation even a few months after the rats had first learned the fear re-sponse. Recently, Melinda Miller, Margaret Altemus, and their collabo-rators reported an as yet unpublished human-based study that suggests propranolol may also impair reconsolidation of conditioned fear respons-es in people who do not have any neuropsychiatric disorder.[4] Trials are now under way to investigate the effects of propranolol on reactivated trauma memories in people who are diagnosed with PTSD.

Another approach is to enhance the process of memory extinction, which in certain ways is the opposite of memory reconsolidation. While reconsolidation updates and strengthens learning, memory extinction re-duces the strength of a memory by repeated exposure to information that conflicts with the original memory. One could also say that extinction is a form of new learning, in which an organism comes to know that a cue originally associated with a traumatic experience no longer precedes trau-ma. For example, if rats are continually exposed to the same tone, but it is no longer paired with a shock, the length of time they freeze in response will progressively shrink. This approach is the basis of exposure thera-py, which is the type of psychotherapy that is most effective for treating PTSD and phobias. People are guided to recall their traumatic experienc-es, or are exposed to phobic cues such as heights, spiders, or public speak-ing under controlled, safe conditions so that they can learn to tolerate the cues without having explosive anxiety reactions or using defense mecha-nisms such as dissociation, which leads to disconnection from reality.

Many researchers are studying the psychological and neurobiologi-cal processes involved in memory extinction, and the possibility of using drugs to facilitate extinction, especially of traumatic memories. Michael Davis, Ph.D., and his colleagues at Emory University showed that Sero-mycin—an antibiotic, also known as cycloserine, used to treat tubercu-losis—can enhance memory extinction in animals.[5] Seromycin activates

NMDA receptors, a type of receptor for the amino acid glutamate, which is known to enhance learning. Davis and colleagues reported this year that they have extended their research to humans, demonstrating that people recover from a fear of heights more quickly if doctors give them Seromycin immediately before exposure therapy sessions. In a separate study by Stefan Hofmann, Ph.D., and his colleagues at Boston University, exposure treatment for anxiety about public speaking was enhanced by giving Seromycin before each psychotherapy session.[6]

Since both propranolol and Seromycin are widely used to treat other medical disorders, one might wonder whether they alter memory when used to treat those disorders. But animal studies suggest that taking the medications repeatedly may not affect memory in the way it does when they are taken in single doses in immediate association with memory recall. Further research is needed.

Another big caveat in assuming these animal memory advances will translate smoothly to human breakthroughs is the communication barrier—we really don't know what animals are thinking. One major problem with animal models of human mental illness is that the core features of the disorders are subjective experiences, rather than observable behaviors. In the case of fear memories, for example, researchers study freezing, startle, and approach/avoidance behaviors in animals. Although they can observe these behaviors, they have no way to access the emotional experience of the animals. When a propranolol-treated rat shows a reduction in freezing behavior in response to a tone that had been previously paired with shock, some scientists interpret that response as meaning that the rat has forgotten the memory that the tone is followed by a shock. But at this point we have no way of knowing whether the memory is actually erased, whether it has merely become inaccessible, or whether only the fear associated with the memory is reduced.

Where Should the Limits Be?

Memory has a fundamental role in human life; in some ways, it defines us. As the President's Council on Bioethics, in its report *Beyond*

Therapy: Biotechnology and the Pursuit of Happiness, points out: "Memory is central to human flourishing . . . because we pursue happiness in time, as time-bound beings. . . . If we are to flourish as ourselves, we must do so without abandoning or forgetting who we are or once were."[7]

The members of the council acknowledged that some memories, such as traumatic memories of violence, war, or disaster, constrain and distort our human experience. People with PTSD and phobias are disabled by their symptoms. The extreme fear and arousal associated with remembering a trauma makes it difficult for people with PTSD to integrate these memories into the rest of their experience and to react in a deliberate, intentional way. Women who have been sexually assaulted, for example, often avoid all relationships with men. In addition, because some victims dissociate, distancing themselves from the situation when danger cues trigger trauma memories, they may lose their ability to properly evaluate dangerous situations. This can make them vulnerable to further assault. A person who developed PTSD after surviving a terrible car accident may react to the sound of a car braking with a racing heart, sweating, and inability to take a step for several minutes. Such problems eventually cause many people with PTSD to avoid situations that might bring back the memory of trauma, sometimes to the extent that they become housebound. Traumatized war veterans often isolate themselves, both to prevent their irritability and hyperarousal from disrupting social interactions and because their hypervigilance makes them experience innocuous situations as threatening.

A treatment that reduced the associated fear could get them out of the house and out into the world, helping them gather the new experiences and memories that are part of the lifelong process of defining and developing a strong sense of self and identity. In the case of a phobia, such as excessive fear of heights or snakes, it is hard to see how lessening a reaction of explosive fear would harm the person with the phobia or change his identity.

Although current research is focused on alleviating emotional and physiological aspects of traumatic memories, such as hyperarousal, we do not know how interfering with the emotional coloring of memory would affect the way we remember and thus how we relate to our past. The potential for relieving human suffering is great, but these pharmacological "tools of interference" with human memories could also be misused. Once memory-blunting drugs are readily available, they may be used not only by trauma victims but also by offenders who cause trauma, as well by witnesses to crimes or accidents. The authors of *Beyond Therapy* hypothesize that blunting emotional responses in those who have committed crimes may also result in diminishing any feeling of guilt, "the psychic pain that should accompany their commission of cruel, brutal and shameful deeds."

At this point, we do not know what aspects of memory can be altered by the drug treatments now being developed for PTSD and other anxiety disorders. Ideally, the disabling fear and autonomic arousal will be reduced, but the memory of a traumatic event might remain intact, and perhaps even become clearer, since intense emotions disturb cognitive processes and impair recall. When patients with PTSD undergo psychotherapy, their memory of the traumatic event often improves. If the treatments also reduce their fear and dissociation, they will have the opportunity to create more complex, and more integrated, personal responses to the trauma, such as sadness, anger, outrage, or remorse. Carefully focused future research will continue to help answer questions about the effects of drug treatments; researchers and society as a whole have this window of time to consider the ethical issues.

If it is possible to erase essential aspects of a memory, the cure for one person may become a way to escape justice and responsibility for others. Before we start using drugs to treat traumatic memories, we must address many ethical as well as scientific questions, a task that can be accomplished only through meticulous research and an intense multidisciplinary dialogue.

CHAPTER 10

Elephants That Paint, Birds That Make Music

Do Animals Have an Aesthetic Sense?

by Lesley J. Rogers, D.Phil., D.Sc.,
and Gisela Kaplan, Ph.D.

Lesley J. Rogers, D.Phil., D.Sc., is professor of neuroscience and animal behavior and founder of the Research Centre for Neuroscience and Animal Behaviour at the University of New England, Australia. She is the author or coauthor of fourteen books, and is an elected fellow of the Australian Academy of Science, and past president of the Australian Society for the Study of Animal Behavior.

Gisela Kaplan, Ph.D., is research professor in animal behavior at the Research Centre for Neuroscience and Animal Behaviour at the University of New England, Australia. She holds two Ph.D.s, one in the arts and another in animal behavior and veterinary science. Professor Kaplan has written eighteen books, some jointly with Lesley Rogers. She can be reached at gkaplan@une.edu.au.

SOME FORTY YEARS AGO, the first gallery exhibition of paintings not of but by chimpanzees shocked the art world and precipitated much debate. The animals had produced abstract paintings pleasing to the human eye. Did this mean they had an aesthetic sense, an appreciation of beauty? Elephants, too, can paint—sales of their canvases are now being used to raise money for zoos and conservation—and so can seals and several other species. Is this really art, or are the paintings more or less accidentally pleasing to us but not to the animal itself? How can we decide whether these strokes of paint are art or mere daubing, made without awareness or any degree of artistic motivation or aesthetic sense? A similar question can be asked about other forms of art, especially music. Birdsong, for example, may be music to our ears, but do the birds appreciate it as an art form?

If research were to prove that animals have an aesthetic sense, we could gain valuable insights into the animals' level of awareness. Creation and appreciation of art are aspects of consciousness that we have traditionally viewed as purely human activities, ones that express our highest cognitive abilities. If animals share at least some aspects of this ability, we will have to look upon them with more respect and perhaps change the ways we treat them. Research on animal art involves studying how the brain perceives sensory information and how we decide whether something is beautiful or has symbolic meaning. Studies in this area also stem from curiosity about the evolution of artistic expression. Looking at the similarities between the art of early humans and that of some primates causes us to wonder if art may have origins that extend back in evolutionary time to the apes, or even earlier.

Mainstream science has yet to be convinced that animals have an aesthetic sense, but these days some scientists who study animals are increasingly convinced that they do have higher cognitive abilities. At the moment, interest is focused on the abilities of animals to solve problems, use tools, and communicate in meaningful ways, but some researchers have dared to suggest that animals may play because they find it pleasurable to do so. Doing something for pleasure, rather than for survival, is part of how we define the act of creating art. But just as we must be open to the controversial idea that animals can create art, we must also be careful of the pitfalls in reaching conclusions too soon.

What Do Animals See When They Paint?

Scientists who study animal behavior have learned that many animals, from fish to apes, invent new patterns of behavior, as did the first Japanese macaque that washed potatoes to remove the dirt before eating them, and that others, especially birds and mammals, behave in ways that depend on forming and using mental representations of both their physical environment and their social context. Some species, it seems, use symbols; others communicate intentionally, for example using specific vocalizations to refer to specific predators. All of these abilities lay a basis for the claim that animals possess consciousness, but they do not prove that the animal is capable of both creating "artistic" productions and appreciating them as mentally and aesthetically pleasing or conveying a symbolic meaning. When we provide an elephant in a zoo or a chimpanzee in a research facility with a brush and canvas, are the paintings they produce art to them, as well as to us?

The first step in deciding whether an animal might have produced a painting as art is to find out exactly what that animal can see. If an animal seems to use color aesthetically but either lacks color vision entirely or is able to perceive only some colors, we would have to conclude that any aesthetic use of color is accidental, however pleasing it may appear to us.

Most paintings by elephants, for example, involve the use of several colors applied in strokes to the canvas with either a brush or the trunk. Individual elephants have immediately recognizable styles, which may reflect each elephant's stereotypical patterns of trunk movement. It is not surprising that elephants are adept at using a brush, since, in captivity at least, they use many different tools.

But elephants cannot see the same range of colors that we do. Recently, Shozo Yokoyama, Ph.D., and colleagues at Emory University measured the visual pigments in the photoreceptor cells of the elephant retina and found that they have only two pigments, compared to our three (we have red, green, and blue cones). Hence elephants are like certain color-blind people, called deuteranopes, who lack one visual pigment and, consequently, see a smaller range of colors than most people do. No behavioral tests have yet been made of the color vision of elephants, but we know that color-blind people, with similar eyes, detect only two primary colors (blue and yellow) and do not see intermediate colors. When blue and yellow are mixed, these people see white or gray, or one of the two basic hues. Humans with normal color vision see four primary colors (blue, green, yellow, and red) and a range of intermediate colors.

Elephants evolved this two-pigment (dichromatic) color vision because they are active during both day and night. In order to see well under both conditions, they traded off some color vision for better vision in low levels of light (at night in moonlight and at dawn and dusk). Being active in the daytime and at night is also true of dogs (at least in their original natural environment), and they also have only two visual pigments, allowing them to distinguish bluish-violet colors from yellowish-red colors but not the range of colors between these two. To judge the artworks of such species, we have to dramatically reduce the range of colors that we see, which, in our opinion, considerably reduces the artistic quality. Color painting seems to be an inappropriate form of expression for animals with limited color vision.

Dogs (and elephants too) can see movement well and might prefer to express art—if that is what they do—in moving pictures. We also must remember that the dog's eyes see well at a distance but cannot focus on close objects. Anything closer to them than about a foot to a foot and a half—as paintings on canvas made by holding a brush between their teeth would be—is out of focus. They use their sense of smell to recognize objects and other animals, so any purely visual representation would lack an essential quality.

Seals in captivity have also been trained to paint, and they use colors too (see examples at www.eagleandowl.com/artan/). But seals are completely color-blind, since they have only one color pigment, green cones, in the cells of their retina. The same is true of whales and the related dolphins, also painters in some zoos. (Since whales and seals are not related species, their monochromatic vision is likely to have evolved for life in the sea, but it is a puzzle why they have green and not blue cones, given that the latter would allow better vision in the open ocean.) Any claim that these species see the colors in the works of art they produce is, therefore, false. This raises a thorny point, because paintings by seals and dolphins are very similar to those by species that can see some color, such as elephants. Perhaps even paintings by species with two color pigments are made without the animals' paying any attention to the colors they use.

Other species see the same range of colors that we do. Primates that are active during the day, such as chimpanzees, are one example. Still other species see an even greater range of colors than we do. Because most birds have four visual pigments, we can only begin to imagine their color-rich world. Ravens have been trained to paint using a brush held in their beak—a Russian raven named Voron and his paintings can be seen at http://animalsart.ru/raven.htm—and they can certainly see all of the colors that they apply to the canvas.

Some species of birds also behave in ways indicating that they possess consciousness. Ravens follow the direction of another bird's gaze, or even that of a human, to see what might be of interest, and they can solve complex problems. One species of raven, resident in New Caledonia, not only uses tools to probe notches in trees for insects but makes the tools; they use their beaks to cut probes from the leaves of pandanus palms. In fact, painting by tame ravens probably depends on this ability to use tools and so is an extension of their adaptation for survival in the wild. Given this evidence of intelligent behavior, we should keep an open mind about the ability of birds to appreciate art.

Symbolism

Do the elephants, seals, and other animals that have been trained to paint use these paintings to represent anything in a symbolic way? None of the works depict anything that we can recognize easily, if at all. The only way that we can answer this question is to ask an animal to tell us what it has drawn. Obviously, to do so we must turn to animals that have been taught to communicate using sign language or by pointing to symbols that signify words. The very fact that apes can learn to communicate with us in these ways shows they have the ability to use abstract symbols.

If signing apes can tell us what they have drawn or painted and if the picture shows any hint of the object, or emotion, that they say it is, we might be convinced that they have indeed created a representation. At least some such examples exist. The chimpanzee Moja, raised and taught to sign by Beatrix Gardner, Ph.D., and Allen Gardner, Ph.D., sketched what she said was a bird, and it did show a likeness, with a body and wings. You can see this drawing at www.awionline.org/pubs/quarterly/su02/moja.htm. Moja used the same schemata when she drew birds on subsequent occasions.

Koko, the famous sign language–trained gorilla, painted what she said was a bird, and it too looked like a body with wings (although perhaps too many wings). We know that Koko, who was able to communicate what she had painted, is capable of abstract thought, because she signs meaningfully about states of mind and behavior (for example, feeling "mad," "hurt," "sad"). Another language-trained gorilla, Michael, has used color symbolically. He was given a variety of colored paints and often painted in color, but he chose to use only black and white to paint what he called "Apple chase," a representation of his black-and-white dog named Apple. Examples of paintings by Koko and Michael can be seen at www.koko.org/friends/kokomart_art.koko.html.

Human art is produced for pleasure. It seems that painting may be pleasurable to animals as well, because animals in zoos often reduce behavior that indicates stress, such as repetitive swaying and self-mutilation, when they are taught to paint. This could, of course, be the result of receiving extra attention from humans, rather than pleasure in the act of painting or in appreciating the painting produced. But chimpanzees may well obtain pleasure from looking at their artworks. Zoologist Desmond Morris, D.Phil., observed one chimpanzee that would scream with what appeared to be rage and frustration if he was interrupted before he had finished his picture.

If drawing materials are available, young chimpanzees will start to scribble spontaneously, without receiving any food rewards, at around one to two years old—about the same age when human infants begin to scribble. The scribbles of the chimpanzee continue to develop complexity, as do those of the human child, but the end point is different. Most chimpanzees stop developing their drawings at a point when the artwork looks rather formless to us, whereas the human child goes on to make easily recognizable representations.

The drawing of the bird by Moja can be compared to the cave paintings of early humans. It is by no means as detailed or accurate a representation of an animal as the Paleolithic cave paintings of bears, bison, antelopes, and so on, but some cave paintings are almost as sketchy, as in the case of ungulates depicted on the Réseau Clastres chamber of the Niaux Cave in Southern France, as well as engravings in Gabillou Cave, also in Southern France (which can be seen in *The Mind in the Cave: Consciousness and the Origins of Art*, by David Lewis-Williams [Thames & Hudson, 2002]). We might, therefore, see the chimpanzee's art as a precursor to that of early humans, and if more examples come to light, we might be forced to push back the origins of art to a much earlier time than currently believed.

Art to the "Artists"

So far we have discussed only painting by animals that are living under artificial conditions, but we can also find examples of what might be called art in the natural behavior of some species. Bowerbirds, for example, arrange objects of selected shapes and colors on their bower as a means of attracting a partner. They have been seen to arrange and rearrange these trinkets, suggesting to some human observers that they may be doing so to meet their own artistic taste. The satin bowerbird also paints the inside walls of its bower with pigments made from a mixture of plant extracts and saliva.

Like the bowerbird's decorated home, paintings by elephants, seals, dogs, ravens, and other species have no recognizable connections to depictions of reality and hence no known symbolism. They are art to us in the sense of modern art, abstract expressionism, but we are far from knowing whether they are art to the "artists." It would be unwise to jump to a conclusion, but the growing evidence of complex behavior and higher cognitive abilities in a wide range of species was unexpected no more than twenty years ago. We should examine the evidence for art with a critical eye, but we should not reflexively close the door to unexpected discoveries.

Is Birdsong Music or Speech?

For humans, music is another essential art. Birdsong is certainly music to us, but it is a matter of debate whether the songs that birds or other animals produce are music to them. In the late 1960s, researchers began to accept that there could be a continuum of cognitive abilities from mammals to humans and to seek evidence of what animals could do. But birds, which were thought to be inferior to mammals and primates, did not fit into this equation. Only recently have scientists begun to realize that many bird species are highly intelligent and may perhaps be aware of the musical qualities of their own songs.

The study of song was not always the province of neurobiology. Up to World War II, songbirds (passerines) were studied in music departments. Researchers were interested in the birds' musical lexicon, their tonal encoding, interval, and rhythm, and how they remembered crucial aspects of sound, either for production of song or for discrimination in listening to it. Oddly, the budgerigar (a small parrot rather than a songbird) was often used as a model species, and that tradition has not completely died. More recent papers describe the budgerigar's astonishing abilities to discriminate musical characteristics of sound (formants, sine waves, timbre, harmonics, and even quarter tones) and to remember these over long periods of time. Judging by the high degree of accuracy and memory shown in budgerigars, it is even possible that they have perfect pitch.

But musicologists are no longer the only ones who study birdsong. Through neurobiological research beginning in the 1970s, evidence mounted that songbirds are capable of a cognitive process known as vocal learning, which depends on auditory feedback mechanisms that can store sounds and commit them to memory. This special ability is quite rare, having been identified in only songbirds, parrots, and hummingbirds, as well as cetaceans, bats, and humans. The main vocal repertoire of songbirds is expressed in song, but their communication system also includes other vocalizations, such as food, distress, and alarm calls. The range of these songs and calls varies markedly among orders and species.

These two approaches to the study of song—musical and neurobiological—create a quandary. The neurobiological tradition equates song with speech (vocal learning), while musicologists regard bird vocalizations as song, belonging to music. In research, speech is tied to cognitive ability, whereas music is linked with creativity. Human beings can learn language, but not all humans possess musical ability.

Learning to Sing

The question of how, when, and to what extent song is learned has been investigated from many different angles. In a classic study, Peter Marler, Ph.D., taught juvenile white-crowned sparrows to sing by having them listen to playback of tape-recorded song. He demonstrated that such learning was limited to the first fifty days of a sparrow's life, which established the concept of a sensitive period in song learning and inspired other investigators to conduct further research on the importance of this time window for development.

Songbirds engage in complex serial learning not only of their own songs but also of their neighbors' (and competitors') songs, and some birds can learn complete songs even if they are exposed to only snippets of information, or phrase-pairs. Yet when a control group was exposed to all the elements of their species-specific song, but each of those elements was presented singly rather than in phrase-pairs, the birds failed to develop normal, full song. In other words, songbirds not only learn; they also use the information that they encounter creatively. Sound learning is also multidimensional—social interaction during the sensitive period is often required for normal song development. For instance, in 1993 Patrice Ad-ret, Ph.D., at the University of Chicago, demonstrated that showing one male zebra finch the head of another male on a video screen roused the experimental bird to learn song and produce it. The type of tutor can play a decisive role in shaping song, and juvenile birds also appear to choose from whom they will learn. For example, zebra finches prefer to be tutored by their fathers, rather than by another adult male, and they prefer more-aggressive tutors. Researchers have also found that pairing auditory and visual cues enhances song learning in nightingales, leading to more reproduction of the song and a larger song repertoire.

Putting all this together, we conclude that the quality of a song that is learned depends on the quality of tuition, on practice, on multi-model presentation, and on social environment. Many birdsongs are crystallized after they are learned, and the perfected song, at least in male breeding song, may be very stereotyped. In this sense, birdsong would probably not qualify as creative and thus could not be called art. The most common human notion of creativity demands that an individual create something new, something unique, for it to be considered art.

Reinvention, Improvisation, and Play

Many bird species, however, improvise and keep reinventing their song, reinvigorating it with new elements, phrases, and sequences. New syllables and phrases, even new repertoire, may be produced in each successive season, as is the case among nightingales and canaries. The brown thrasher is thought to hold the record, at close to 2,000 song types. Nightingales organize the elements of their songs into hierarchies and follow rules of how the songs are constructed, similar to the way humans use syntax. In addition, each individual bird invents its own songs and so creates aspects of singing (new phrases or "sentences"), which can be used to identify the individual bird. Some birds continue to change their repertoire throughout life and, in a few extreme cases (as in the brown thrasher), may never ever repeat the same song.

Because scientists who study animal behavior, including song, traditionally search for its function, none of these highly variable songs or the changes in repertoire of the best singers have ever been considered "creative" or "art." Since most research has focused on species in which only the male sings—and he does so only during the breeding season—song has been said to serve the bird's purpose of holding territory and competing against other males to secure a female for mating.

However, in some species singing does not seem to be associated with territoriality or reproduction. Some birds simply sing to themselves when they are alone. This behavior does not seem to fit the assumption that all animal behavior must serve a function that aids survival, leading us to wonder if such singing could be a form of leisure activity or play, which would bring us closer to the idea of creativity, of music for music's sake. This kind of singing has been observed repeatedly. Irene Pepperberg, Ph.D., who has studied the vocal productions of African Grey parrots for decades, noted that the hand-raised parrots engage in sound play, most often when they are alone but sometimes when humans are present. The sound play can include both mimicking human speech and making parrot sounds.

Gisela Kaplan's research on the Australian magpie has also found that a major proportion of singing in magpies occurs when their offspring have grown up and territorial defense is not of immediate concern. Indeed, some of the most beautiful song by magpies comes when the bird is alone and self-expression is at its peak. At times this song is skillfully embellished with mimicked sequences and phrases, which we call cadenzas in music. Some of Kaplan's recorded magpie songs certainly can be described in musical terms—the bird's voice moves across four octaves, varies its phrasing between staccato and legato, and embellishes the sequence with vibrato, trills, or deep overtones. Moreover, when a song is complete, an individual bird will end it with a closing phrase all its own. It sings this signature phrase in much the same way that painters put their names or initials on completed paintings.

Such playful, even creative, singing—particularly if it is not connected to reproduction or territorial functions—is ignored by many researchers, but it is celebrated by others. David Rothenberg's book *Why Birds*

Sing: A Journey Through the Mystery of Bird Song (Basic Books, 2005) identifies some birdsong as evidence of creativity, and the author, himself a musician, says his own creativity has been inspired by birdsong. In a book edited by Nils Wallin, Björn Merker, and Steven Brown and titled *The Origins of Music* (MIT Press, 1999), the ever-changing song of the humpback whale is described as music and as evidence of a creative process, rather than as constrained by function. Whether we should label as music the changing songs of birds and whales and call the process creativity (that is, art) is still a matter of conjecture. Such performances have no name in mainstream science today, but scientists are starting to push against the barriers.

Implications of Animals as Artists

At the current state of scientific knowledge, we can say that many species of birds and mammals have much more complex ways of behaving than were thought possible even a decade ago. Animals certainly can be trained to produce paintings that we may wish to call art, and we have some evidence that apes, at least, draw images that to some degree match what they tell us they represent. How extensive these abilities are across species and whether they occur in animals in the natural environment remain unknown. We suggest that to avoid making mistaken judgments, we should consider each species separately, taking into account what it sees and hears as well as its ability to perform other functions that we associate with consciousness.

Does it matter whether animals have an aesthetic sense or may be motivated to create art? And if animals do have an aesthetic sense and produce art, are there any implications for research, for our scientific theories, or for the way we treat them? Because scientists have traditionally assumed that the ability to create and enjoy art does not exist in animals, researchers still know next to nothing about what such an ability might be like. But we would answer all of these questions with a cautious yes.

First of all, scientific theories about animal behavior would have to be changed. Human creativity and art are generally associated with leisure, not with fulfillment of basic survival needs, and the ability to create art is related to general arguments about cognitive ability. So, if animals are shown to have an aesthetic sense, we might have to step outside the scientific framework that seeks survival value in all aspects of an animal's behavior and that draws a clear line between the capabilities of human brains and those of other species.

Moreover, we can see implications for theories about the origins of art, because aspects of artistic expression may have been present much earlier than the evolution of modern humans. Finally, animal welfare paradigms could be affected. For example, we might realize that sounds and colors matter as much as structures in the way housing for animals is organized, whether in zoos, research facilities, or other human settings, and that we should have a much broader perspective on the types of activities we make available to these animals. Ultimately, finding that some animals share a sense of aesthetics—as humans use the term—might well change our sensitivities and attitudes to animals overall, offering further evidence to dismantle the outworn claim that animals are "just" animals.

CHAPTER II

Transforming Drug Development Through Brain Imaging

by Paul M. Matthews, M.D., D.Phil., FRCP

Paul M. Matthews, M.D., D.Phil., FRCP, is vice president for imaging, genetics, and neurology in clinical pharmacology and discovery medicine at GlaxoSmithKline (GSK) and head of the GSK Clinical Imaging Centre, Hammersmith Hospital, London. He also is professor of clinical neurosciences at Imperial College, London and (honorary) MRC clinical research professor at the University of Oxford. He is the co-author, with Jeffrey McQuain, of *The Bard on the Brain: Understanding the Mind Through the Art of Shakespeare and the Science of Brain Imaging* (Dana Press, 2003).

ACCORDING TO THE EIGHTEENTH-CENTURY French philosopher Voltaire, "The art of medicine consists in amusing the patient while nature cures the disease." Even as late as the beginning of the early twentieth century, the great Oxford physician Sir William Osler wrote, "One of the first duties of the physician is to educate the masses not to take medicine."

The subsequent years of the twentieth century saw a transformation of these pessimistic views. The discovery of penicillin was the most dramatic example of a "magic bullet" for acute, life-threatening infectious disease. However, although the stories are less well known, neurology paved the way for effective drug therapies for chronic disease. Success in alleviating the symptoms of neurological conditions has come in the use of drugs for treatment of epilepsy, Parkinson's disease, migraine, pain, multiple sclerosis, amyotrophic lateral sclerosis, and even Alzheimer's disease. With effective treatments to offer, neurologists have become more than simply spectators to the disorders of brain and mind. They are able to intervene with a scientifically informed expectation that the course of a disease can be changed.

But more than half a century since the modern age of pharmaceuticals was ushered in with the "magic bullet" of penicillin, no major affliction of the brain and mind has a cure. Many, such as dementia, depression, schizophrenia, and stroke, remain without satisfactory treatment. As our population ages, the burden of disorders of later life, frequently neurological, becomes ever greater. At least one in four of us will develop Alzheimer's disease if we live to the age of 85. Stroke is the third leading cause of death and the major cause of chronic later-life disability in the United States. As our population grows overall, the cost to society in lost work days and lost human potential will increase.

What is keeping us from finding powerful new drugs to relieve patients with these major disorders? Most people in the pharmaceutical industry now estimate that typically $1 billion or more must be spent to develop a compound, test it in the laboratory and then in clinical trials, and finally obtain approval for it as a new drug. Development of a new drug typically takes fifteen or more years from identification of a promising target to approval and sale of the drug. (To learn more about this arduous process, read *The Long, Sometimes Bumpy Road of Drug Development,* p. 164.) The number of hurdles that must be met is so great that, in recent history, only one out of ten compounds that entered initial human trials make it to market. For molecules targeting diseases of the brain, the record with conventional approaches has been even less encouraging, with perhaps only one out of 100 chemicals proposed as potential drugs receiving approval for sale. Moreover, many of the diseases for which treatments are now being sought have chronic courses too long to fit into conventional drug development timelines. Many are complex disorders involving interactions between several genetic, developmental, and environmental factors, so that drugs targeting any one process may be expected to have only a partial effect on the overall course of the disease.

The Potential of Brain Imaging

To help experimental medicine drive the development of drugs in humans faster and with greater confidence, one especially promising area for innovation lies in the use of new technology. Brain imaging methods that allow scientists to watch the living brain in action, non-invasively, are among the most promising of the new technologies. Used in exploratory ways, new imaging methods can better track the activity of diseases, providing more-sensitive measures of patient characteristics than is possible with usual clinical observations. Researchers can directly probe molecular interactions by using imaging to observe responses in patients' brains, which will permit better selection of the right dose of experimental drugs. Information can be enhanced, and the effect of the drug being studied can begin to be predicted by measuring drug effects on brain functions that are relevant to the disease.

Modern imaging that allows scientists to watch the functioning brain now relies primarily on two approaches. Magnetic resonance imaging (MRI) uses harmless magnetic fields and radio waves to map brain structures and brain physiology. Positron emission tomography (PET) uses safe, tracer doses of radioactive materials to follow the fate of individual molecules as they travel in the human body. These techniques promise to change the way drugs are developed. Future drug development will spend less time studying "models" of disease in animals and move quickly to more-informative experimental medicine in humans. The benefits of using neuroimaging in this way can already be glimpsed in six aspects of drug testing: time and cost, confidence in targets, integration of information, dosage, drug combination, and understanding of the placebo effect. The FDA and other regulatory agencies are encouraging efforts to develop and validate these kinds of approaches.

Smaller, Faster Clinical Trials

Imaging tools can enhance the sensitivity of measurements of patient responses to individual drugs. Consider the example of Alzheimer's disease. Because the clinical symptoms progress slowly, researchers using conventional clinical measures of the disease must study either very large numbers of patients or a smaller group over very long periods of time in order to have significant results. Both alternatives make clinical trials expensive and slow to complete.

However, elegant studies from the laboratory of Eric Reiman, M.D., in Arizona, among others, have shown that even early stages of Alzheimer's disease are associated with reduction in the brain's use of glucose, a natural fuel for the brain.[1] More surprisingly, Reiman and colleagues have demonstrated that reductions in glucose utilization seen in a PET scan may precede noticeable symptoms of Alzheimer's disease even by decades and that these reductions can be measured with high precision.[2] Calculations suggest that use of a PET outcome measure could decrease the number of patients required for early clinical trials by more than tenfold. While still expensive, PET scanners that can track glucose utilization in the brain are now widely available in major medical centers, making this a feasible approach for new trials.

Loss of brain volume is associated with the degeneration of nerve cells in Alzheimer's. What is remarkable now is that using MRI scans, which provide extremely sensitive measures of brain structure, small changes in brain volume can be measured with extraordinarily high precision: out of the approximately one-and-a-third-quart brain, changes in volume as small as half a teaspoon can be seen. This allows the brain atrophy associated with Alzheimer's disease to be plotted over periods as short as six months, using populations much smaller than those needed if researchers look only at conventional clinical outcomes.[3]

Increasing Confidence in Targets

With MRI measures, early clinical trials for some types of Alzheimer's disease treatments could become more efficient, but how can possible new targets identified with genetic searches or other methods be validated as likely prospects? No simple, general strategy exists. One popular approach is to use transgenic technology to create animals that make either too much or too little of the gene product in question, looking for a change that would correspond to one associated with the human disease. But how does one assess dementia in a mouse? Poorly!

A promising new approach is to find specific measures of activity in the human brain that are informative about malfunctions related to a disease and then to look for changes in these brain functions that are associated with the slightly different forms of genes that are found between people.[4] This approach uses normal human variation to identify mechanisms of disease.

The laboratory of Daniel R. Weinberger, M.D., of the National Institutes of Health, has pioneered this type of approach for psychiatric diseases. In an illustrative study, his group used functional magnetic resonance imaging (fMRI) to study brain signals in people who had taken the stimulant drug amphetamine.[5] They learned that differences in the signals were associated with small differences in the structure of an enzyme responsible for inactivating neurotransmitters in the brain (COMT, or catechol-O-methyl transferase). Paradoxically, rather than improving thinking and memory, subjects with the more rare form of COMT experienced impaired performance after taking amphetamine. The use of modern imaging to relate genetic differences to drug response differences between people provides an efficient approach to pharmacogenetics—the tailoring of drugs to people who are more likely to respond well to them based on their personal genotype. This type of study also provides clues to genetic features that may predispose a person to addictive behavior or depression.

Integrating Information

Combining information from more than one imaging tool can provide further insights by relating changes at a molecular level to differences in the way large systems in the brain function. For example, researchers have directed considerable effort toward developing drugs that may help to reduce the liability to or severity of addictions, such as addiction to alcohol. The neurotransmitter dopamine (the same small molecule that when reduced in movement areas of the brain causes voluntary motion to freeze in Parkinson's disease) is also implicated in this disorder of motivation.

Work by a research group in Cologne, Germany, illustrates this particular benefit of neuroimaging.[6] They took a group of alcoholics, who had been abstinent from alcohol, and assessed the craving that they expressed when shown attractively displayed pictures of beer. As expected, the abstinent drinkers all expressed a desire for the beer (which they suppressed), but different individuals had different degrees of craving. The investigators then measured the concentration of the large molecule in the brain that allows dopamine to signal to neurons (or more precisely, to bind to specific receptors on the neurons) in a region known as the ventral striatum, which is a key element in the brain circuit controlling motivation. They found that the higher the craving, the lower the amount of the receptor that was able to bind a dopamine-like PET tracer. This demonstrated an association between abnormalities of dopamine and malfunction of its receptor.

The German group went on to use fMRI to probe the relationship of this specific biological effect to broader physiological responses in the brain. By alternately presenting pictures of beer or neutral pictures, they identified differences in brain activity in response to the alcohol cue, finding that these differences highlighted the same parts of the brain identified in the PET experiment. The researchers also observed a direct relationship between the magnitude of the fMRI signal and the binding of the PET probe to the dopamine receptor. Together, these observations further confirm a central role for the dopamine system in craving behavior, validating it as a target for drug therapy. While animal experiments have suggested this, here the information is coming not only from human studies but, more important, from one of the specific human diseases of interest for a possible new drug. Greater confidence in the validity of targets should limit wasted effort in drug development on targets that are not functionally important.

Getting the Dose Right

Once researchers have identified a validated target for a drug, they can develop molecules to interact with the target. How much of a potential drug must be given in order to have the desired effect? This is a particularly critical question, because too much drug could be harmful by leading to unwanted interactions or overly inhibiting an important pathway. Even with marketed drugs, ensuring that each patient gets the optimal dose remains a major problem. Consider anti-epileptic drugs. If an insufficient dose is administered, seizures may continue unabated; if too high a dose is given, the patient may experience unacceptable slowness of thinking or other neurological symptoms, such as blurring of vision or unsteadiness.

One clever way of addressing this problem is to develop a PET tracer molecule that sticks to the drug target if there is no real drug present but is knocked off when the drug is around. By measuring the amount of the PET tracer sticking to the target in people taking different doses of the real drug, it is possible to determine exactly the right amount of

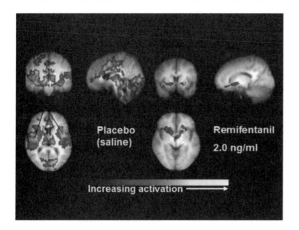

Figure 1   Positron emission tomography (PET) images the distribution of a safe, tracer dose of a radioactively-labelled molecule injected into the body. This example shows where a drug that is intended to treat a neurological disease binds to have its action. The image shows strong binding in deep regions of the brain, confirming that it reaches the intended site of action. This type of information can be of critical importance early in drug development. *Credit: Courtesy of Prof. Irene Tracey and Dr. Richard Wise, FMRIB Centre, University of Oxford.*

drug needed to get the desired result. The higher the dose of the real drug, the lower the amount of the tracer that binds to the target (because more drug is bound). This all can be explored non-invasively, very early in drug development (even in Phase I trials, which test drugs for safety in healthy individuals). This will allow a rational choice of the dose used in a Phase II clinical trial to be defined with confidence.

An alternative new approach now generating much interest is the use of fMRI to assess brain responses to different doses of a potential drug. By looking at the relationship between brain response and drug dose, researchers can estimate the clinically effective dose. Recent studies have shown, for example, how emotional areas of the brain (in the limbic system) respond to anxiety-provoking fearful faces presented in an fMRI examination.[7] Administration of an anti-anxiety drug such as Valium will reduce this response. Because fMRI responses can be measured relatively quickly, various doses can be tested over a short period of time, potentially allowing the selection of an effective dose after study of only a small group of healthy volunteers or patients.

Combining Drugs

Increasingly, doctors will treat major symptoms or diseases by using two or more drugs together. This allows each drug to be used at doses that minimize unwanted adverse effects, while combining to produce an additive therapeutic effect. In the brain, we can find many examples of how several functional systems act together to generate a particular behavior, symptom, or disease. Since we know too little about the use of this strategy for Alzheimer's disease to use that as an example, let us consider instead the problem of treating pain.

Most broadly, pain is not simply a sensory phenomenon; it also involves an emotional response to a stimulus. Bad pain is not just strong but also deeply unpleasant and disturbing, or even frightening. In addition, the perceived intensity of pain or its unpleasantness is determined by our expectations of the pain. All nurses and general practitioners are well aware of this reaction—for example, when they administer injections to children, they try to do so quickly, before the child is aware of the needle coming.

Studies using fMRI have allowed the complexity of pain to be probed in increasing detail. In one experiment, led by my Oxford University colleague Irene Tracey, D.Phil., some years ago, either pleasantly warm or painfully hot thermal stimuli were applied to the back of the hand soon after a particular colored light was turned on—one color for the pleasant sensation and another for the painful one.[8] Tracey and her colleagues identified sensory, emotional, and attention-focusing regions of the brain that were more active during the periods associated with the burn than with the gentle warmth. Together, these regions form the parts of the brain that create the perception of pain. All of the fMRI signals related to this perception could be turned off if a powerful pain-relieving drug, such as morphine-like remifentanil, was given. The extent to which the signal was suppressed was related directly to the dose of the drug.[9]

Tracey and her colleagues were able to extend their observations beyond simply understanding pain by investigating what went on in the seconds just before the painful heat or the gentle warmth was applied, when the study participants viewed the colored light that signaled what was to come. The fMRI response during this period provided a brain measure of anticipation. As we might expect, the period before the painful heat was associated with real anxiety—even the most hardened research volunteer does not like being burned. Tracey found that distinct areas in the front part of the brain were active during this time. Use of an anti-anxiety drug such as Valium could block the signal from these regions selectively. This result illustrates how imaging can provide measures of pharmacodynamics (a technical term for the response to a drug), which can be more informative than conventional measures because of the selectivity and precision with which they can be measured. By better understanding precisely which aspects of pain processing are affected by different drugs, it is possible to more rationally think of ways in which drugs can be used together to get maximal benefit.

Understanding the Placebo Effect

A particular complexity in developing treatments for brain disorders lies in the way in which activity of the brain can change its own responses. We often observe a strong placebo effect (a benefit from inactive agents) in disorders of the mind. Recall Hamlet's words: "Nothing is either good or bad but thinking makes it so."

The placebo effect can be thought of as a response changed simply by expectation. In a drug treatment trial, this response complicates efforts to determine the true effectiveness of a drug. Even if neither the study participants nor the investigator knows which group of subjects is getting the active drug and which is getting an inactive comparator

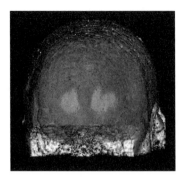

Figure 2   Functional magnetic resonance imaging (fMRI) can be used to define drug effects. Here the brain response to a painful heat stimulus is shown during infusion of saline placebo or after administration of remifentanil, a morphine-like analgesic. After the remifentanil, most of the brain activity signaling pain during the heat stimulus disappears, along with the subject's perception of pain. *Credit: Courtesy of Dr. Tony Gee, GSK Clinical Imaging Centre, UK.*

(placebo), the placebo effect tends to reduce the potential magnitude of difference between responses from the two groups.

A frontier area for brain imaging, particularly in pain studies, lies in trying to establish which brain systems are involved in placebo effects so that we can determine whether markers exist that could distinguish placebo responses from drug responses.[10] Finding such markers, particularly if they could be applied easily to a larger population, could result in substantial reductions in the size of populations needed to test drugs, enhancing the speed and reducing the cost of the tests.

New Tools, New Potential

What will the future bring for imaging technology in drug development? A frontier area for partnerships between engineers, doctors, and pharmacologists lies in finding ways of translating sophisticated, expensive methods such as fMRI and PET into cheap, portable tools that can be used in large trials in all regions of the world, including the developing world. One example already in place is optical coherence tomography (OCT). OCT can be performed with a relatively inexpensive desktop device that uses a low-energy beam of laser light to rapidly scan across the retina of a subject looking into a computer-driven optical scanner. From sensitive measurements of differences in the way that reflected laser light bounces off the retina, a tomograph, or picture, of the back of the eye can be generated that allows the thickness of the retinal nerve layer and its distribution across the retina to be mapped with exquisite precision.

OCT was developed originally to assess nerve degeneration in diseases such as glaucoma, in which fluid pressure in the eye is increased, causing degeneration of the optic nerve. However, it also can be used for brain diseases. For example, David Miller, M.D., and his colleagues at University College, London, have recently shown how the degeneration of the optic nerve in multiple sclerosis can measured sensitively.[11] This advance presages a potential for the technique to be used to assess new drugs that might limit nerve degeneration in multiple sclerosis, which is thought to be the major cause of progression to full disability in that terrible disease.

New imaging technologies that allow direct study of the human brain and human brain disease promise to make drug development for diseases of the brain and mind faster and cheaper, delivering both more drugs and more-effective ones. However, we must also be cautious in the use of this technology. What we want to treat is the disease—the *dis-ease* of the patient—and that ultimately can be assessed only from the clinical effects—the symptoms and the outcome. Imaging is a tool to help move toward this goal with greater confidence, not a replacement for careful clinical observation of the patient.

With modern imaging, genetics, genomics, and information technologies we should now be in a position to rethink current drug development paradigms and ask, Can we move from the stage of having a new target to a new drug in just five years? Can we dramatically cut the high cost of drug development (which benefits no one)? Can we search for cures, rather than contenting ourselves with symptom management? The challenges of brain diseases are enormous. But with new imaging tools and an increasing understanding of brain function, there is hope that—even within our generation—new approaches to drug development will allow us to progress toward realizing real brain health all through life.

The Long, Sometimes Bumpy Road of Drug Development

by Paul M. Matthews, M.D., D.Phil., FRCP

WE OFTEN READ THAT understanding the human genome promises the transformation of medicine, because finding the genes associated with major diseases provides potential new targets for therapy. So what are the obstacles in developing new drugs?

To understand this formidable challenge, let us consider the example of creating a new drug to treat Alzheimer's disease. Scientists long knew that a protein called amyloid formed "plaques" in the brains of people with Alzheimer's but not whether it was involved in causing the disease. Identifying abnormalities in genes regulating the synthesis of beta-amyloid in the brains of patients with an inherited form of Alzheimer's helped remove the doubt in most scientists' minds. But how can this help us develop a new treatment?

Studies with isolated cells in a culture showed that beta-amyloid can kill cultured nerve cells grown outside of the body. Together with the information from genetics research, this suggests that the progression of Alzheimer's might be controlled (or, in early stages, reversed) by clearing amyloid from the brain or blocking its production.[1] Different strategies for doing this have been proposed, such as finding ways of "washing" beta-amyloid from the brain, using large molecules called antibodies that irreversibly bind to beta-amyloid. Such therapeutic antibodies could be either injected (passive immunization) or generated through vaccination. Just imagine a future in which children are given a quick jab to prevent mumps, measles, rubella, whooping cough—and Alzheimer's disease. Another drug strategy would be to develop chemicals that inhibit the brain enzymes (large molecules made by cells in the body that perform particular chemical tasks) that are responsible for producing beta-amyloid in the brain.

Having established a biological strategy and rationale for the treatment of Alzheimer's disease, the drug industry faces the next challenge: to engineer molecules that can perform the required tasks. This means engineering an antibody (or other binding molecule) that can potentially pull beta-amyloid out of deposits in the brain or designing a chemical that can block the actions of enzymes that produce beta-amyloid. The rapidly growing predictive power of computational biology makes designing such molecules increasingly scientific, but important elements of the process remain an art. Many, many antibodies must be produced and screened in order to discover the few that work well, or researchers must synthesize a variety of small molecules expected to bind to the enzyme target. Luck continues to play a role, and not all biologically validated targets can be further developed to become potential drugs.

The next hurdle is to show that a molecule can do what is intended in a test tube or in cells in a culture dish, such as selectively recognizing beta-amyloid or blocking the target enzyme without interfering with other enzymes that have important functions. And successful test tube experiments are not enough. Researchers must show that the potential drug can be administered safely to an animal and, ideally, that it can reduce beta-amyloid in an animal model of Alzheimer's disease.

Many molecules appear in the test tube or even in animal models to function in ways that suggest that they could be drugs, but either they fail to have the right effect in humans or they cause unacceptable side effects. So the most challenging hurdle for drug development is testing a molecule in humans. The first question is whether a safe dose range can be determined, one that will allow high enough doses to have the action predicted in the earlier animal experiments. For instance, in our example of a hypothetical drug for Alzheimer's, researchers would have to establish that the beta-amyloid antibody can be given at high enough doses to bind significant amounts of beta-amyloid while not triggering undesirable activation of the body's immune responses. Or they would need to show that an enzyme-inhibiting small molecule does not damage the liver, which is responsible for deactivating many small molecules in the bloodstream. This kind of critical safety information is acquired in Phase I experimental trials, in which the drug is carefully administered to closely monitored healthy volunteers. Success in Phase I is not guaranteed; overall, 35 percent of candidate drugs fail here for one reason or another.

But even finding that potential drugs work in an animal model and are safe in humans does not mean that they can be used to treat a disease. Candidate drugs that survive Phase I trials must next be tested in Phase II clinical studies in patients. In our example, Phase II trials would test whether the new molecules are likely to be effective in Alzheimer's disease. At this point researchers encounter a whole new set of challenges.

Consider how this would work with the much more straightforward problem of developing an antibacterial drug. Phase II trials might involve testing whether administration of the compound leads to faster reduction of fever or other signs of infection. This can be done relatively quickly, because the trial uses clinical measures and signs that are easy to interpret; the fever goes down, for example, or swelling is reduced. But with neurological disorders such as Alzheimer's disease, which have time courses extending over years or decades, it can be difficult to assess the potential utility of a possible drug, even in a preliminary way, if only conventional clinical measures are used.

To appreciate this problem, consider that in late 2005 more than 30 new potential drugs for Alzheimer's disease were being developed by different companies. Using even the most sensitive current clinical measures of memory function and cognition, and even assuming that any of the compounds achieved as much as a 50 percent slowing of rates of cognitive decline, investigators would need more than 300 patients in order to make a preliminary assessment of the efficacy of each one of these agents after the first year of follow-up.[2] In fact, establishing studies, recruiting patients, and analysis take far longer. What this means is that a traditionally designed Phase II study would take at least a couple of years for Alzheimer's disease—and more than one

Phase II trial is typically needed because different populations must be studied and a range of information must be acquired. Just do the math: 300 patients per trial x years per trial x more than one trial per possible drug. And only about a third of the new molecules entering Phase II can progress to Phase III.

Completion of Phase II with proof that a molecule might have useful activity in treatment of a disease still does not make a drug. Larger, longer Phase III studies must follow in which the drug is carefully tested in a more usual treatment population. The goal would be to assess not only whether the new potential drug is effective but also whether it is better than existing treatments. Equally important is understanding any risks associated with taking the new candidate drug. For Alzheimer's disease, this might involve recruiting a larger number of patients from several different medical centers. Just under two-thirds of the molecules entering Phase III will be developed further. After Phase III testing, approval for marketing of a drug finally can be sought from government regulators. Then, and only then, can the drug reach the person with Alzheimer's who needs it so desperately.

CHAPTER **12**

Hardwired for Happiness

by Silvia Helena Cardoso, Ph.D.

Silvia Helena Cardoso, Ph.D., a behavioral biologist and science writer, is founder of the Teleneuroscience Center and founder and director of the Edumed Institute, University of Campinas, Brazil.

THE PURSUIT OF HAPPINESS is the engine that moves human-kind, that motivates us to study, to work, to marry and have children, to make friends, to pursue all sorts of worldly pleasures, to dream of the future, and sometimes to fight for social and financial status. Of all the goals we may pursue in life, happiness is the only one to have worth in itself; all the others—health, power, money, beauty, success—make sense only as means of achieving it. To many people, life would be unbearable without the belief that they can be happy.

Are our brains "hardwired" for happiness? That is, does happiness have a biological basis, rooted in the evolution of the nervous system? If so, understanding how our brain machinery works to make us happy could suggest ways to transform our behavior and our relationships with others, as well as to set up a better society.

Of course, to define happiness objectively is difficult. Joseph LeDoux, Ph.D., the well-known researcher and author on the physiology of emo-tions, expressed the scientist's frustrations with this elusive entity well when he wrote: "There are many answers for what are emotions/happiness. Many of them are surprisingly unclear and ill-defined." Neuroscientist Richard J. Davidson, Ph.D., observed that the word *happiness* "is a kind of a placeholder for a constellation of positive emotional states. Of all the emotions, happiness is the one scientists least understand."

Studying the many enemies of happiness, such as stress, depression, anxiety, and phobias, is essential, of course, but recently researchers have turned their attention to the brains of happy people as well. Even though scientists still do not agree on the precise definition of happiness, they are beginning to discover a vast array of biological counterparts to what we associate with the idea of happiness. Functional imaging methods and the study of key neurotransmitters in the mechanisms of emotion in the brain and body have led not only to a deeper understanding of the biological components of happiness but also to practical applications for achieving it, such as antidepressant drugs.

Searching for Happiness in the Brain

Neuroscience, biology, and psychology all have important roles in deciphering and elucidating the mechanisms and purpose of positive emotions. Great scientists, from Charles Darwin to William James to Sigmund Freud, have studied in detail our most basic negative emotional processes, such as fear, stress, anxiety, anger, and aggression, and how they relate to the brain, nervous system, hormones, and internal organs. Their findings provided most of the knowledge we have today about the neural correlates of emotion in general, particularly the role of subcortical structures such as the limbic system, hypothalamus, thalamus, basal ganglia, and midbrain.

Positive emotions, however, used to be considered too subjective and difficult to study, so for a long time neuroscientists neglected them. Unhappiness was considered to arrive on its own, since fear, anger, and defense are responses to danger from the external world and are vital for our survival ("fight or flight"). But our feelings of pleasure and happiness were thought to be largely cultural and were regarded only as guiding our behavior toward desirable situations.

In the middle of twentieth century, however, new approaches and techniques for investigating the nervous system began to provide more rigorous ways to evaluate positive affective (that is, mood) states. The first real breakthrough came in the 1950s, when American psychologists James Olds, Ph.D., and Peter Milner, Ph.D., discovered what they named the "pleasure centers in the brain."[1] Rats that were implanted with electrodes in certain areas of the brain learned to press a lever that would deliver pulses of electrical stimulation to these areas. Most of the experimental animals were so taken by the stimulation effects that they would not even stop to eat, drink, or rest, but would press the lever at fantastic rates until they were totally exhausted. They would even cross electrified grids to reach the lever.

The phenomenon of self-stimulation of the brain occurred only in certain parts of the subcortical brain, not in others, showing that specific structures processed motivational input; the nucleus accumbens was

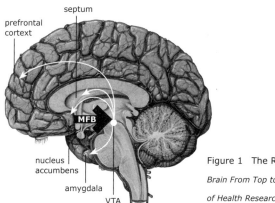

Figure 1   The Reward Circuit. *Credit: The Brain From Top to Bottom, Canadian Institutes of Health Research.*

one of the most active. Stimulation of the nucleus accumbens in humans elicits smiling, laughter, pleasurable feelings, happiness, even euphoria. Extensive mapping proved the existence of a coherent "reward system" in the brain, which was also shown to exist in many mammals, including humans. As shown in Figure 1, when the cortex has received and processed a sensory stimulus indicating a reward, it sends a signal to the ventral tegmental area (VTA) in the midbrain. The VTA then releases dopamine not only into the nucleus accumbens, but also into the septum, the amygdala, and the prefrontal cortex. These regions are connected through the medial forebrain bundle (MFB).

What is the secret to the behavior elicited by stimulation of the nucleus accumbens? The answer is dopamine, which is involved in functions ranging from motivation and reward to feeding and drug addiction. Dopamine is a neurotransmitter, a chemical substance released by neurons at their synaptic connections to other neurons in the brain. The first neurotransmitter to be associated with positive emotions and feelings, it is essential for activation of the reward system because it sets in motion the neural circuits involved in motivation. The dopamine-driven reward system is best known for its association with addiction, in which it causes uncontrollable urges to engage in a destructive behavior.

According to neuroscientist Kay Jamison, Ph.D., author of *Exuberance: The Passion for Life* (Vintage Press, 2005), dopamine transmission in the brain may also be related to the exuberant temperament, as well as to the mania, of bipolar disorder (also known as manic-depression). When asked whether there is a clear dividing line between exuberance and mania, Jamison said, "Exuberance can escalate into mania in people who are predisposed to manic-depressive or bipolar illness. Most exuberant people never become manic, but those who have bipolar illness often have an exuberant temperament."

A second breakthrough in understanding the neural basis of positive emotions came in the 1970s, when American neuroscientists Solomon H. Snyder, M.D., and Candace Pert, Ph.D., discovered that our brains produce endorphins, a kind of internal morphine composed of a sequence of amino acids. Receptors for endorphins, called opiate receptors, can be found in several parts of the brain. When released by the pituitary gland and by neurons in the hypothalamus, endorphins suppress pain. In addition, pleasurable feelings that accompany actions such as eating chocolate, laughing, smiling, touching, meditating, singing, listening to good music, and even orgasm are partially attributed to the brain's release of endorphins.

Endorphins released in the brain also increase the release of dopamine. As proposed by neuroscientist Kent Berridge, Ph.D., at the University of Michigan, wanting (desire) and liking (pleasure) appear to be two distinct biological processes with separate but interrelated neurochemical systems in the brain, both related to positive emotions, including happiness.[2] Although many gaps in the scientific knowledge remain to be filled, Pert went on to propose that opiate receptors and endorphins provide a biomolecular basis for emotion and are the key to the effect of emotions on our health.[3]

What Brain Imaging Shows

The fantastic progress in techniques for obtaining functional images of the brain—color images that reveal precisely what brain structures are activated and deactivated when certain emotions or behaviors occur in human beings—has had a major impact on neuropsychology in the last decade. These images allow us to study the brain basis of emotions in a non-invasive way, without having to intervene in the brain as scientists do with experimental animals, for example by implanting electrodes or creating lesions.

In the 1990s, Antonio Damasio, M.D., Ph.D., discovered that positive and negative feelings are both generated and processed by different parts of the human brain. Damasio and his group at the University of Iowa were among the first to use positron emission tomography (PET) to map the brain correlates of emotions, both negative and positive.[4] The researchers, who took PET scans of brain activation while volunteers made themselves feel anger, fear, sadness, and happiness, made several interesting discoveries. First, different emotions activate or deactivate different areas of the brain. Happiness, for instance, had a functional pattern remarkably distinct from sadness, sometimes in opposite ways. Second, the activation and deactivation patterns exhibited marked asymmetries; that is, the two sides of the brain reacted differently to the induced emotions. Happiness activated the right posterior cingular gyrus, as well as the left insula and the right secondary sensorimotor cortex. Sadness, as one would expect, decreased activation in these regions. Other structures in the basal region of the brain, such as the pons, were activated in sadness but not in happiness.

Damasio proposed a distinction between emotions and feelings. According to his "somatic marker hypothesis," the sensory system detects peripheral changes in the heart, circulatory system, skin, and muscles that are commanded by emotions in the brain and interprets these changes as feelings.[5] This could explain why Aristotle considered that the heart was the seat of the soul and emotions, something we still see in our everyday language and symbols—for example, the universal sign of love is a heart. As various biochemicals act on the neural circuits of our hearts, we discern different patterns and strengths of heartbeats, which lead us to feel, variously, contentment, happiness, love, joy, despair, depression, fear, or anger. The heaviness of heart we feel in an amorous deception is quite different from the flutter of passion.

The brain processes involved in voluntary control of negative emotions, which enables normal, healthy people to resist sadness and depression, were demonstrated by a group of neuroscientists from Montreal, Canada.[6] They studied functional brain images of people induced to feel sadness by watching short films and those of people who were able to suppress this feeling by an internal effort. The study showed that sadness caused changes in the right ventrolateral prefrontal cortex, the anterior temporopolar cortex, the affective division of the cingular cortex, and the insula (all regions that have been previously associated with human emotional regulation). On the other hand, the voluntary suppression of sadness activated the right dorsolateral prefrontal cortex and the right orbitofrontal cortex.

Note the role of the right hemisphere in connection with negative feelings, and the role of the prefrontal cortex in processing basic emotions. The ventrolateral prefrontal cortex in particular seems to be part of the circuits processing information from the body when activated by emotion. So it might be involved in the somatic marker system proposed by Damasio. The role of the dorsolateral prefrontal cortex in suppressing sadness that was observed in the Montreal study agrees with previous studies showing that it is also involved in the willed suppression of positive emotions, sexual arousal, and other feelings. This part of the brain seems to be related to holding information in temporary memory. We might speculate whether the right dorsolateral prefrontal cortex could be trained to suppress negative emotions and thus make people happier.

An explosion of imaging studies of emotions followed the development of functional magnetic resonance imaging (fMRI), which is easier to use than PET. For many, the abundance of results seemed to conflict with each other and muddled considerably any effort to find a single cohesive interpretation. A recent survey of 106 studies,[7] for example, could not determine a concordance of activated areas for happiness and sadness that was as clear-cut as the initial studies by Damasio. Other areas that appeared regularly in most of the imaging studies of happiness were the rostral supracallosal anterior cingulate cortex and the dorsomedial prefrontal cortex. Many studies have associated the anterior cingulate cortex with the regulation of emotions, so it has been named "the affective division" of the cingulate cortex. It also appears to be altered in people with depression, for whom happiness is difficult to achieve.

Also, Richard J. Davidson, Ph.D., and his group at the University of Wisconsin reported activation of the left frontal part of the brain while study participants watched happy video clips, as well as when Buddhists were meditating.[8] Many studies have observed important functional differences between the left and right sides of the prefrontal cortex. The left side seems in general to be associated with positive emotions (lesions in this side of the brain cause depression), while the right side is associated with negative emotions.

Most of the studies reported so far seem to support a theory of emotions that applies to happiness: the "central affect program." Although first proposed by Charles Darwin in his 1872 book *The Expression of the Emotions in Man and Animals*, the current description of this theory has been championed by noted emotion researchers such as Paul Ekman, Ph.D.,[9] and Jaak Panksepp, Ph.D.[10] According to this theory, an affect program is a brain mechanism that stores patterns for and triggers complex stereotyped emotional responses, which are present in the same form in all humans and cultures. Specific affect programs are controlled by interconnected brain structures and develop over time.

Recent imaging studies have given credence to this theory, because the activation of discrete brain regions has been correlated with specific emotions, such as activation of the amygdala by fear. A clearer definition of happiness and better research methods, particularly for inducing the feeling of happiness under laboratory conditions, are necessary, but functional neuroimaging holds great potential for non-invasive studies of the happy brain in operation.

The Set Point for Happiness

Happiness is both a general state of being and the result of specific time-delimited events. Psychologists have established that each person has an average overall level of happiness at any particular period of life.

When you ask someone whether he is happy, he usually answers quickly and with assurance, reflecting his appraisal of the average during a relatively recent period. This average state, or baseline, has been defined by researchers such as David Lykken, Ph.D., as a "set point" of happiness,[11] in the same way that a stable level of glucose concentration in the blood is set by the body or the temperature is set for a refrigerator.

Different people have different set points of happiness. Discrete events, such as the day you marry or have your first child, or when one of your parents dies or you are fired from your job, cause a sudden temporary increase or decrease in your level of happiness. But most people almost invariably return to their set points at some time after the especially happy or unhappy event. Moreover, the set point is normally above neutral—most people lean more toward being happy than unhappy—and unhappy events have less influence and are more quickly forgotten than happy ones.

The general set point of happiness can be modified downward by chronic disturbances, such as depression, or upward by medication. Some antidepressant drugs, for example the serotonin reuptake inhibitor fluoxetine (Prozac), actually seem to be able to alter the set point in some people, leading to such drugs' being dubbed "personality cosmetics" and "happy pills." Cognitive behavioral therapy may also have this power.

Just as one cannot be unhappy all the time and be considered healthy, one cannot be euphoric (*euphoria* means an excess of happiness) or exuberant all the time (*exuberant* means excessively enthusiastic). A healthy person will soon return to normal, previous levels of happiness. In fact, excessive, out-of-context euphoria and exuberance are hallmarks of pathological conditions, such as hypomania and the manic phase of bipolar disorder. Kay Jamison observes, "In their mild forms exuberant states are intoxicating and adaptive but in their extremes they are pathological; in short, exuberance can range from imagination and exploration to recklessness and madness."

The observation that happiness is a fluctuating state with a set point suggests the hypothesis that it is regulated internally, as are other basic organic and mental states, by homeostatic mechanisms, which control dynamic adjustments to maintain a stable condition. As the saying goes, "Time is the best healer." In other words, the homeostatic mechanism returning one to one's normal level of happiness eventually prevails.

Evidence from Evolution

All anatomical and physiological characteristics of human beings were molded by natural selection. Emotions are primitive components of human behavior that are processed by older parts of our brain, such as the limbic system, the hypothalamus, and the brain stem. According to Norwegian biologist Bjorn Grinde, D.Sc., D.Phil., the human capacity for positive and negative feelings was shaped by the forces of evolution, so the evolutionary perspective should be relevant to the study of happiness. The evolutionary perspective has four important correlates.

First, because we share basic emotions and their neural substrates with other mammals and non-human primates, in theory a predecessor for happiness should exist in animals. Some ethologists agree that chimpanzee behavior, observed both in the wild and in captivity, suggests that an internal state analogous to human happiness can be found in these animals.

Second, happiness should be represented in the brain in the form of hardwired circuits; otherwise, it could not be selected during evolution. We have examined some of the evidence that points to this, and I think that eventually objective neuroscientific studies will confirm that this is exactly the case.

Third, a set of genes and a mechanism for genetic expression are needed in order to construct specific brain circuits. David Lykken (who first proposed the idea of a happiness set point) was also responsible for research that has provided the strongest evidence so far for a genetic basis for happiness. In a study of identical twins separated at birth, some 60 percent of the likelihood that each twin would describe himself or herself as happy was accounted for by common genetic factors, not by environmental differences in their lives. Lykken argues that "the laws governing happiness were designed not for our psychological well-being but for our genes' long-term survival prospects."

Finally, to support the evolutionary perspective, happiness should have a direct or indirect value for the survival of its controlling genes. Because happiness is a positive emotion, we can hypothesize that it acts as a good motivator and internal reinforcer of behavior, particularly for achieving long-range goals that are important for the survival of the organism or species. One of the leading researchers on the brain structures related to motivation, Larry Swanson, Ph.D., has found growing evidence that the brain is hardwired for happiness via goal-seeking behavior.[12] He says: "Setting and achieving goals can have an amazing influence on creating well-being in our daily lives and making us feel happy. This is because the steps involved in goal-directed activity, namely motivation, goal seeking, successful outcome, and feelings of pleasure, are wired into the brain's structure."

A behavioral chain composed of tens of thousands of individual acts (such as those necessary, for instance, for graduating with a medical degree many years after you initially wanted to become a doctor) must be somehow reinforced along the way, and happiness is a good candidate to be that reinforcer. A hunting, bipedal hominid would need not only motivation but continual reinforcement in order to plan and execute the long chain of behaviors that would lead eventually to his presentation of a carcass of meat to his family for their survival. So every step, such as preparing the weapons, running through the hunting grounds, planning the kill, and felling the prey, would generate some combination of enthusiasm, determination, hope, joy, and contentment, all components of what we call happiness.

Going further, many psychologists have proposed a hierarchy of needs. Some needs are more important than others, but happiness is associated with fulfillment at all levels. The best-known schema, created by Abraham Maslow, Ph.D., is set up as a conceptual pyramid [Figure 2]. Needs must be satisfied in order of priority: first the physiological needs (food, water, rest) and safety (shelter, protection against enemies and dangers); then love

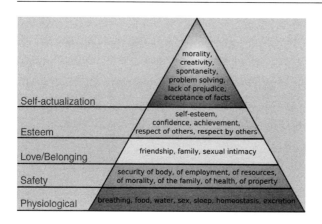

Figure 2   Maslow's Hierarchy of Needs. *Credit: J. Finkelstein/ Wikipedia.*

and belongingness (for example, the search for positive affect, social participation); esteem (self-esteem and the search for the respect of others); and, finally, self-fulfillment (self-sufficiency, vitality, creativity, meaningfulness, and so on).

As with other primitive emotional components of our minds, such as fear or anger, over the course of evolution happiness has become a dominating feature of many aspects of our life. While neuroscientists have not yet agreed on all of the critical brain regions and processes that underlie it, happiness seems to be at least partly determined by hardwired, genetically expressed structures in our brains. For human beings, happiness is a complex mix of nature (hardware) and nurture (software), greatly influenced and modified by cultural experiences and learning.

Toward a Definition of Happiness

Drawing together what science has learned about the biological basis for positive emotions such as happiness, we can begin to move toward a definition against which future research discoveries can be measured. I suggest that happiness involves:

- a general average level of contentment and well-being, with a high frequency of positive feelings such as good humor, joy, laughter, hope, and enthusiasm, coupled with relative freedom from negative feelings such as sadness, worry, anxiety, anger, irritability, despondency, and despair;

- the presence of more positive (happy) than negative (unhappy) events in our lives, and, more important, the ability, when negative feelings occur, to allow them only a minimal effect on our emotions (and on our bodies as well);

- a personality (both genetically and culturally determined), an individual disposition, and behavioral traits that make a person more resilient to adversity and more prone to enthusiasm, laughter, and good humor;

- a match between our life expectations and our deeds. Aristotle wrote that "happiness is the consequence of a deed"—that is, it is the result not of chance but of using for the best all the opportunities that we encounter in our lives.

Helping Happiness Prevail

People need to be happy in order to build a satisfying world and thrive in it. Our own nature will then reward us with its best—inner peace, pleasure, and joy. But unfortunately what human nature gives, it also takes back rather quickly. After our moments of bliss, back we go to our previous level of happiness, our set point. This oscillation appears to be the way of life, but a more enduring state of happiness is possible if we persist in helping our positive emotions to prevail.

Science has shown again and again that being happy is good for your mind and your body. Happy people are more confident, optimistic, energetic, and sociable. They are also better prepared to deal with difficult situations, are more enjoyable to live and to work with, and have a higher capacity for pursuing their aims and acquiring the means to achieve them. In addition, happy people appear to be healthier and live longer. For instance, just recently Sheldon Cohen, Ph.D., of Carnegie Mellon University published a study confirming his earlier discovery that people who are happy or exhibit other positive emotions are less likely to become ill when they are exposed to a cold virus and, when they do catch a cold, they have fewer symptoms.

For much of humankind, getting enough sustenance for physiological needs, survival, and safety is a constant battle, which many, unfortunately, lose. Satisfying even these most basic needs requires the joint work of emotions and cognition, of subcortical as well as neocortical brain systems, all acting under the command of our prefrontal brain, the jewel of human evolution. Those of us who are more fortunate, however, can seek to reach other levels of Maslow's pyramid of human needs, such as achieving self-esteem, self-actualization, authenticity, and meaningfulness. Such continuous growth seems to be an important ingredient of a happy life.

Research in psychology has shown repeatedly that the ability to regulate one's emotions is essential for a happy life. While we may strive to be rational and in control, emotions are an indissoluble and essential part of our psyche. The great personal search, then, is how to defeat our inner enemies, to achieve control over our negative emotions. Although many psychologists and neuroscientists decry as unsupported sensationalism what has been known as the "power of positive thinking," in fact several serious studies, using functional brain imaging techniques to observe the brain during sadness and happiness, have shown that distinct parts of the prefrontal cortex are involved in the volitional suppression of negative feelings.

The effectiveness of cognitive behavioral therapy in dissipating automatic negative thoughts and feelings has demonstrated that achieving happiness by self-control is not impossible. Relaxation and meditation techniques, originally developed by philosophies and religions of the Far East, have also proved helpful in increasing positive emotions and controlling negative ones. Neuroimaging research has shown that meditation is accompanied by a clear increase of activity in the left prefrontal cortex, known to be related to positive emotions, and a decrease of activity in the parietal cortex, related to spatial localization (thus facilitating the meditator's becoming more concentrated on his inner self). Equally effective for many people are techniques used to activate the brain's reward systems and increase levels of dopamine, serotonin, and endorphins through pleasurable sensations such as music, light, color, and touch.

Of course, we should not forget that mind power is not enough. Individual propensity for happiness also depends on our genetic heritage (more than 60 percent, according to studies with twins). Both innate temperament and negative early experiences in life, such as traumatic stress and abuse, are extremely influential. Fortunately, however, we can end this article on an optimistic note. Although we cannot change our genetic makeup or our individual past, science has developed many ways of correcting and even healing detrimental influences of those factors on happiness. And we are certain that the future will provide an even greater and more impressive set of techniques and tools to help us look inside the happy brain and, as a result, discover more ways in which people can claim their own happiness.

CHAPTER 13

Hope for "Comatose" Patients A *Cerebrum* Classic

Nicholas D. Schiff, M.D., *and* Joseph J. Fins, M.D.

Nicholas D. Schiff, M.D., is an assistant professor of neurology and neuroscience at the Weill Medical College of Cornell University and assistant attending neurologist at New York Presbyterian Hospital. Dr Schiff directs Weill's Laboratory of Cognitive Neuromodulation, where he conducts research on the neurophysiological mechanisms of arousal and forebrain integration and clinical studies of the pathophysiology of impaired consciousness. He can be reached at nds2001@med.cornell.edu.

Joseph J. Fins, M.D., is chief of the Division of Medical Ethics and professor of medicine, public health, and medicine in psychiatry at Weill Medical College of Cornell Medical Center. Dr. Fins is director of medical ethics at the Cornell campus of New York Presbyterian Hospital. His research interests are in end-of-life care and the ethics of neuropsychiatric research. He can be reached at jjfins@med.cornell.edu.

From Volume 5, Number 4, Fall 2003

ON JULY 11, 2003, newspaper headlines proclaimed the dramatic awakening of Terry Wallis, a thirty-nine-year-old Arkansas father who had been in a "coma" after suffering a head injury in a July 1984 car accident. He had been riding with a friend when their car plunged into a creek. When they were found under a bridge the next day, his friend was dead and Wallis was comatose. But now, nineteen years later, he was talking. His first words were "Mom" and then "Pepsi," and, over the ensuing weeks, he began to speak with greater fluency. He apparently had no memories of the intervening time. In his world, Ronald Reagan was still president. The media described his recovery as a "miracle," and his doctors were stunned. What occurred seemed scientifically beyond the realm of possibility.

Despite the very unexpected (and as yet unexplained) nature of what happened to Wallis, we were, perhaps, less surprised than many people. For a decade we have been conducting research at the frontiers of understanding impaired consciousness and the ethical challenges posed by devastating brain injury. We have seen other patients like Wallis, whose improvements, although less heralded, also defy our understanding of impaired consciousness that follows brain injury. Our goal has been to understand both the mechanisms of recovery and biological differences between those patients who remain forever unconscious after catastrophic injury and those who regain at least limited awareness.

From our own research and that of a handful of other cognitive neuroscientists, we knew that the media's portrayal of Wallis's condition before he recovered was inaccurate, at best, and, at worst, seriously misleading. Although he was portrayed as being in an irreversible coma, or in a vegetative state, Wallis was in neither. He had not been in a coma immediately before his recovery, because "coma" describes a state of unconsciousness typically lasting only weeks from the time of injury. Comatose patients usually either recover or slip into various longer-term states of impaired consciousness. A review of literature about Wallis indicates that his behavior during the nineteen years after the accident was also inconsistent with being in a vegetative state. He had been able to respond to simple questions with a nod of the head or with grunting sounds, indicating some level of awareness and interaction with his environment—neither is seen in the vegetative state. These behaviors, noted before his dramatic recovery, suggest a state that scientists are only beginning to characterize: the minimally conscious state.[1]

The importance of these distinctions cannot be overstated. Identifying those patients with severe brain injuries who have a chance of recovering is the first step in deciding who may benefit from the therapeutic approaches now being developed that might help them to regain function and independence. Having said that, we come face-to-face with a puzzling paradox: why has this progress in understanding impaired states of consciousness been met by a surprising lack of interest—not to say an attitude of dismissal—on the part of the scientific community and society at large?

Our Hidden Epidemic of Traumatic Brain Injury

At the outset, we should appreciate that what happened to Wallis when he was twenty years old is an all too common story. Although most of us do not think about traumatic brain injury (TBI) until a family member is touched by its tragedy, its incidence is staggering. TBI is the leading cause of long-term disability in children and young adults and, in the United States alone, has 1.5 to 2.0 million victims a year. Motorcycle, automobile, and sporting accidents are among the most frequent causes (Figures 1 and 2). The toll of TBI is still more graphically demonstrated when we consider that head trauma has left between 2.5 and 6.5 million people in the United States with some degree of permanent impairment. The yearly cost for new cases of TBI is between $9 billion and $10 billion, and lifetime costs per individual have been estimated to be between $600,000 and $1,875,000.

Even these numbers, however, fail to do justice to the burden of TBI. Lives are suddenly and irrevocably altered by severe head trauma. If a patient is lucky enough to survive the acute phase of injury and intensive care, but remains severely impaired, he may face years of rehabilitation on the road to recovery. Unfortunately, rehabilitation services are often

---

THE HIDDEN EPIDEMIC OF TRAUMATIC BRAIN INJURY IN THE UNITED STATES

- 1.5 million TBI injuries occur every year.
  - 50,000 people die from their injuries.
  - 80,000 to 90,000 people experience long-term or lifelong disability as a result of their injuries.
  - 2,000 people enter a persistent vegetative state following their injuries.
- One-third of all injury deaths are the result of a TBI.
- The cost of TBI is estimated to be more than $48 billion each year, including $10 billion for new cases.

THE MOST FREQUENT CAUSES OF TRAUMATIC BRAIN INJURY IN THE UNITED STATES

- Vehicle crashes
  This includes motor vehicles, bicycles, recreational vehicles, and pedestrians.

- Firearms
  Firearm use is the leading cause of death relating to a TBI, and 90 per cent of people with a firearm-related TBI die.

- Falls
  Falls are the leading cause of TBI among the elderly, and 60 per cent of fall-related TBI deaths involve people over 75.

Figures 1 and 2

limited, and third-party payment depends on evidence of the patient's continual improvement and on demonstration of what is called "medical necessity." Some patients move from hospital to rehabilitation facility and then return to a life that is markedly different from their former existence. Such was the experience of Trisha Mellie, whose autobiography, *I Am the Central Park Jogger*, tells the story of a young investment banker attacked while jogging in 1988 and her recovery from that nearly fatal attack. Mellie describes how she had to relearn all the tasks she had once mastered as a child and to develop new strategies to compensate for her loss of cognitive function. Even more daunting, she had to grapple not only with memory loss but with the realization that her injury had changed who she was as a person.

As monumental a challenge as Mellie faced, another class of patients with TBI has sustained even greater impairment. This class includes people such as Wallis, who recover from their acute injuries and have (perhaps minimal) evidence of cognitive awareness but fail to meet the criteria for medical necessity required to qualify for intensive ongoing rehabilitation. Our health-care system fails them and their families. After a brief period of coma rehabilitation—or none at all—they are exiled to nursing homes for what is often impolitely called "custodial care." The experience of Wallis and his family is not unusual. His parents report that, after the accident, Terry was never seen by a neurologist. He was placed in a nursing home, and his family was told that an evaluation would be expensive. His father said to reporters, "They told us it would cost $120,000 just to evaluate him to see if he could be helped, and we didn't have that kind of money." The Wallis family applied for Medicaid to cover the cost of evaluation but was turned down. "They said the government will not put out that kind of money on no more chance than he's got to reenter the workforce," reported his father.

Even if the government was correct that the cost of evaluation would be more than Wallis's potential wages if he recovered, what is our ethical obligation to patients like him? We think society owes the many people with TBI, and their caregivers, some intellectual curiosity about severely impaired consciousness, as well as the potential fruits of scientific investigation. In the light of new understanding of various brain states after injury, clinicians, patients, and their families need accurate information to make informed choices about care. But, sadly, many clinicians are themselves ill informed, so they are unable to discuss the options for treatment—or refusal of it—in a meaningful way. If a society must ration scarce resources, the decision about who to treat should be based on the best available science and an accurate assessment of diagnosis and prognosis. Such clinical precision seems to be the least that physicians should demand of themselves when caring for these patients.

States of Disordered Consciousness: A Primer

What brain states can follow head injury? The immediate consequence of a severe brain injury, like the one that Wallis sustained, is a loss of consciousness that results in a brain state known as coma, an "unarousable unresponsiveness." The person does not respond to vigorous efforts to elicit a response of any kind—sound, movement, or eye-

COMA	PERSISTENT VEGETATIVE STATE (PVS)	MINIMALLY CONSCIOUS STATE (MCS)
Does not respond to vigorous efforts to elicit a response of any kind, shows no variation in behavior.	Demonstrates no awareness of self or response to their surroundings.	Demonstrates directed behaviors and evidence of awareness of self or the environment, cannot demonstrate consistent functional communication.
Appears to be in a sleeplike state with eyes closed.	May demonstrate reflexive behaviors, including occasionally smiling or appearing momentarily to focus his/her gaze on something.	May demonstrate fluctuating behavior, including basic verbalization, gestures, memory, attention, intention, and awareness of self and environment.
Typically lasts only weeks from the time of injury. Can recover, die, or evolve into another state of impaired consciousness.	Can live for years in a PVS and never recover, or could progress to a minimally conscious state or recover.	Can emerge from a MCS and regain full consciousness.

Figure 3

opening—and shows no variation in behavior, simply a sleeplike state with eyes closed.

The prognosis for someone in a coma very much depends on the person's age, the amount of structural damage (as identified by brain imaging), and whether there is evidence of direct injury to the brain stem. From coma, very severe brain injuries can progress to brain death, a total loss of whole brain function, including brain stem activity. In other cases, the comatose state, if uncomplicated by other factors, is typically followed within seven to fourteen days by an indeterminate period during which an eyes-open, "wakeful" appearance alternates with an eyes-closed, "sleep" state. These alternating periods represent a limited recovery of cyclical change in arousal pattern and characterize the vegetative state (VS), as originally defined by Bryan Jennett, M.D., and Fred Plum, M.D., in 1972.[2] In all other respects, the vegetative state is similar to coma. Patients in vegetative states demonstrate no evidence of awareness of self or response to their surroundings. If the patient remains in a vegetative state for more than thirty days, he is deemed to be in a persistent vegetative state (PVS). Prospects for the recovery of consciousness become grim when the vegetative state becomes chronic or permanent, after three months in the case of anoxic injury, caused by oxygen deprivation, and a year following traumatic injuries.

In other cases, a patient may recover to the point of very limited but definitely observable responses to his environment. Such a patient is classified as in a minimally conscious state (MCS). In this condition, a patient exhibits bits of directed behaviors that are different from the reflexive behaviors seen in PVS patients. The difference is that MCS patients demonstrate unequivocal—albeit fluctuating—evidence of awareness of self or the environment. Limited behavior exhibited by MCS patients can include basic verbalization, gestures, memory, attention, intention, and awareness of self and environment.

To know that a patient has emerged from MCS, we must observe consistent functional communication. Crossing this threshold requires more than the ability simply to follow commands. For example, a patient may be able to correctly identify a printed "yes" or "no" on a card held up by an examiner but not be able to answer questions reliably using such signaling. The patient would be considered only at the borderline of emergence from MCS.

In Terry Wallis's case, we see him emerging from MCS after he passed through an initial coma and a period in the vegetative state. Although Wallis has often been described as vegetative right up until he began to speak, in fact he had been able (possibly within the first year after his accident) to respond to simple questions with a nod of his head or grunting—hallmarks of MCS. If so, his pattern of recovery is wholly consistent with scientific understanding of the recovery of consciousness from a vegetative state resulting from traumatic brain injury.

Because the progression from PVS to MCS may take months following a traumatic brain injury, physicians who do not fully understand what is happening and rely solely on their observations of a patient can have unnecessarily negative expectations for the patient's recovery. Information about the patient's underlying brain function, gained from neuroimaging, may change those expectations. New neuroimaging techniques may eventually become an important adjunct to careful neurological examination and lead to much earlier identification of patients, like Wallis, who may be able to emerge from MCS. Just as surely, though, they might tell us that no further recovery can be expected, even if a patient exhibits some limited behavior above a vegetative level. Having more knowledge, sooner, will bring hope to some and despair to others.

Seeing into the PVS Brain

The first brain-imaging studies of PVS patients were done by Fred Plum, M.D., David Levy, M.D., and their colleagues in the early 1980s using a technique called fluorodeoxyglucose positron emission tomography (FDG-PET). This imaging technique measures how much energy the brain is consuming. Plum and Levy discovered that overall brain-tissue metabolism in PVS patients was half or less than half of normal.

On the basis of their work, we can now pose a critical question: what functional activity might remain in severely injured brains? Seeking an answer, together with Plum, we have collaborated with research groups at the New York University Center for Neuromagnetism (directed by Urs Ribary, Ph.D., and Rodolfo Llinas, M.D., Ph.D.), and the Memorial Sloan-Kettering Center (directed by Brad Beattie, Ph.D., and Ron Blasberg, M.D.). To try to size up the remaining functional activity in several PVS patients, we used three neuroimaging techniques: magnetic resonance imaging (MRI), magnetoencephalography, and quantitative PET analysis.[3] What we have discovered is new evidence that the persistently vegetative brain can harbor still-functional, but isolated, networks and that these networks, at times, can generate recognizable fragments of behavior.

We became interested in the possibility of this residual function through the case of a forty-nine-year-old woman who had suffered a series of cerebral hemorrhages as a result of malformed blood vessels in her brain. Despite some two decades in PVS, this woman occasionally uttered single words (typically expletives) without any external stimulus. MRI imaging showed that her right basal ganglia and thalamus were destroyed, and FDG-PET measurements confirmed a marked overall reduction of more than 50 percent in brain-tissue metabolism, which is consistent with what we know of cerebral metabolism in PVS. What was intriguing, though, was that several isolated, relatively small regions in her left hemisphere showed higher levels of metabolism. These regions, in the normal adult brain, are associated with language functions. Two other patients we studied also revealed isolated metabolic activity in the brain that could be correlated with other unusual patterns of behavior. It seems, then, that residual cerebral metabolic activity that remains after severe brain injuries is not random; it is tied to local cerebral networks that have been preserved and to patterns of neuronal activity.

The work of other scientists helps to fill in the picture. Using a different PET technique, David Menon, M.D., and his colleagues in Cambridge, England, found that a patient who was recovering from PVS into MCS had isolated neural networks that responded to human faces. Steven Laureys, M.D., and his colleagues in Belgium have examined functioning in the PVS brain by comparing its responses to simple auditory and other stimuli with its baseline resting state. For both types of stimuli, these PVS patients demonstrated a loss of brain activation in so-called higher-order regions—regions outside of their primary sensory cortices.[4] This seems to indicate that there is a wholesale disconnection between functions in the PVS brain that prevents basic sensory input from being processed anywhere but the earliest cortical levels. This evidence is consistent with our own results and supports the conclusion that residual cortical activity seen in PVS patients does not signify any awareness.

One additional observation may shed light on the significance of residual islands of activity in the PVS brain. The observation involves one patient who had exceptionally widely preserved metabolism in the cerebral cortex, despite six years in a vegetative state after a traffic accident. The patient's behavior had been completely unremarkable. The unusual observation was that this patient's cortical metabolism was near normal, except for marked reductions in the severely damaged region of the upper brain stem and central thalamus. We conjectured that this well-preserved cortical metabolism probably meant that there were many partially functioning brain networks. In other words—and this is crucial—nothing linked these islands of activity as before the injury. It is relevant that the upper brain stem and central thalamus regions were damaged in this patient because those regions have a critical role in the functional integration of parallel neural networks. In a different patient, who had recovered from a vegetative state, Laureys had identified a return of activity in these regions.

Inside the MCS Brain

Everything we had learned about the brains of PVS patients made us want to study what remaining cerebral activity might be found in MCS patients, particularly those recovering almost to the level of emergence from MCS. Patients who remain near this borderline raise different questions. What underlying mechanisms could be limiting their recovery of communication? In collaboration with Joy Hirsch, Ph.D., and her colleagues, we studied two such patients and compared what we found with our findings in PVS patients.[5] Both MCS patients could intermittently follow simple commands with eye movements, occasionally made attempts to vocalize, and showed significant fluctuations in their responses. Would their brains respond to language? To test this, we played them taped narratives, spoken by a familiar relative. Tapes were played both as normal speech and backward.

We found that when the story was played forward, as normal speech, the two MCS patients showed activation of cerebral networks underlying language comprehension. The activation was similar to activation in normal subjects. Not so for the tape played backward. Normal subjects showed similar activation patterns for both, but the MCS patients failed to activate the language comprehension networks when they heard the tape reversed. This failure indicates to us that in some MCS patients there are forebrain networks that might be potentially functional, yet fail to establish the patterns of activity needed for consistent communication. The preservation of forebrain networks associated with higher cognitive functions, such as language, could provide a neurobiological basis for wide fluctuations in behavior, such as was observed in Terry Wallis.

Obviously, these studies suggest the crucial importance of functional integration. Although these MCS patients demonstrated functioning forebrain networks and could respond to forward language, their overall resting cerebral metabolism was low, near PVS levels. To our surprise, we also found that these patients seemed to have intact integrative responses in both cerebral hemispheres. This leads us to believe that differences in the integration of functions, more than levels of resting brain activity, are what separate PVS from MCS.

Emerging from MCS

With this insight about what may be happening in the brains of some patients with severe brain injury, we can revisit the question of how someone like Terry Wallis could harbor residual cognitive capacities that lay dormant for so many years. One possibility is that, over time, as patterns of activation come and go in intact regional networks, one result may be improved awareness and cognition, but another result, exactly the contrary, may be the inhibition of recovery.

From a physiologic standpoint, several mechanisms may possibly be at work here, yielding changes in the capacities of patients who have complex brain injuries. How and to what extent these mechanisms may limit recovery we do not know as yet, but several observations are suggestive. It is relatively common, even after a localized stroke or brain injury, to have reduced cerebral metabolism in brain regions that are remote from the site of injury. The cause seems to be a loss of excitatory inputs from nerves at the site of the original injury. This process, which is reversible, results from a strong inhibition of the distant neurons brought about by a lack of incoming synaptic activity.

Another mechanism that conceivably could affect the delicate balance of excitation and inhibition is abnormally increased synchronization of populations of neurons (such as is seen in epilepsy and other brain disorders). It is possible that, following structural brain injuries, such changes may arise in specific brain networks that play an important role in the functional integration of networks in the normal brain. An alteration of this kind could have played a role for Wallis, limiting his capacity for producing speech through active inhibition of language networks. While this explanation is, of course, speculation, some kind of functionally reversible process must play a role in such cases. Wallis's doctors speculated that adding the antidepressant Paxil to his medications could have been connected in some way with his later recovery of speech, although he had taken Paxil for eighteen months before his recovery. Did this antidepressant have a role in slowly changing his patterns of cerebral integration, and eventually unmasking residual function in his brain?

As neuroimaging is used to study additional patients with severe brain injuries, more questions will arise about the mechanisms underlying their functional disabilities. For patients like Wallis, neuroimaging could also reveal previously unrecognized residual capacities that can be at least partially restored by new kinds of therapy.

Exploring Deep Brain Stimulation

If some patients with severe cognitive impairment could be limited, in part, by a lack of functional integration among intact regions of their brains, we should look for ways to foster reintegration. Patients like Wallis who have recovered to functional levels that are near the threshold of emergence from MCS would be the first likely candidates for new therapies to improve consistent communication. Of many new medical technologies, the most promise may lie in emerging techniques for deep brain stimulation.

Over the past fifteen years, deep brain stimulation has advanced the treatment of drug-resistant Parkinson's disease, sometimes dramatically, and is approved by the Food and Drug Administration for that use. To date, more than 15,000 patients with Parkinson's have been treated and new uses of deep brain stimulation are being investigated to help patients with chronic pain, epilepsy, and psychiatric disorders such as depression and obsessive-compulsive disorder.

In an interdisciplinary project with the Cleveland Clinic Foundation, the JFK-Johnson Rehabilitation Center, and the Columbia University

Functional MRI Research Center, we have been planning how to use deep brain stimulation to raise the functional level of MCS patients. Our efforts follow some provocative work over the past two decades that studied this technique with patients in a vegetative state (including, in fact, Terry Schiavo). Specifically, there have been several attempts to use deep brain stimulation in regions of the central thalamus, an area with many connections to the cerebral cortex. Activating these brain regions with an electrical current induces many of the standard signs of arousal, confirming experiments with animals in which electrical stimulation induced wakeful arousal. Unfortunately, in some fifty PVS patients studied worldwide, the stimulation evoked no evidence of sustained recovery of interactive awareness.

In contrast, deep brain stimulation did succeed in bringing about significant physiologic responses in PVS patients, who had large increases in global and regional cerebral metabolism and changes in brain wave activity toward a more normal profile for a wakeful state. The behavioral and physiologic arousal seen in all the patients demonstrated that despite overwhelming brain damage it was still possible to activate the cortex. It may be that, given the overwhelming brain injury in PVS, this increased activation was not enough to restore interactive awareness.

The areas electrically stimulated in these PVS patients are parts of the thalamus that are known to link a state of arousal with some aspects of moment-to-moment behavior. Here, then, is a new rationale for using deep brain stimulation in MCS patients who demonstrate limited integrative forebrain activity. Unlike the PVS patients—who initiate no behavior, follow no commands, and attempt no communication—MCS patients near the borderline of emergence typically have changes in cognitive functioning that come and go over hours, days, weeks, or even longer. This fluctuation might be the result of unstable interactions of the arousal state with the organization and maintenance of behaviors, which these patients can initiate, but not sustain. If so, then deep brain stimulation of the central thalamus might improve integration in the damaged networks that underlie these limited behaviors. By contrast, functional and structural neuroimaging studies demonstrate these networks in patients with chronic PVS have been overwhelmingly damaged.

Before we can go beyond these planned pilot studies of deep brain stimulation in MCS patients, the criteria for selecting patients must be worked out, and important ethical questions considered. At present, it seems that those who have recovered to functional levels near the threshold of emergence from MCS could be the first candidates for new therapies to improve consistent communication. But new therapies for patients with severely impaired consciousness encounter challenges in the form of attitudes and preconceptions that could pose greater difficulties than the science. In particular, as we work with MCS patients, we will have to address two sources of skepticism: the right-to-die movement and the troubled record of psychosurgery.

Seeking a New Moral Warrant

The first hurdle will be for our society to reexamine its attitudes toward patients with severe brain injury, attitudes shaped by the right-to-die movement. By taking to heart the possibility of new hope for these patients, we are asking for a moral warrant to intervene in patients who resemble those patients for whom the right to die was first established in the 1960s and 1970s. This hard-won—and important—right was vouchsafed to patients closest to death and for whom, therefore, the withdrawal of life-sustaining therapy seemed justifiable: patients with permanent and irreversible loss of cortical function.

Because the futility of any potential treatment was pivotal in justifying the right to die for PVS patients, many physicians remain nihilistic about potential interventions in these patients with severely impaired consciousness. Why bother, they wonder, because these people are essentially dead? In ruminations like these, which underlie judgments but are seldom explicitly voiced, people echo perceptions that were critical in establishing the rights of patients to refuse life-sustaining therapies. In the seminal 1976 Karen Quinlan case, the New Jersey Supreme Court allowed the removal of life-sustaining therapy because Quinlan was in a vegetative state without, the court held, any possibility of return to a "cognitive sapient state." In 1968, a similar justification was urged by

Harvard anesthesiologist Henry K. Beecher, M.D., when he advanced the concept of brain death, although that was in the context of seeking organs for transplantation. In both cases, however, the moral value placed on life and death hinged, in large part, on a person's cognitive state.

We hope, of course, to intervene for patients who are in the minimally conscious state, not the hopeless condition of chronic PVS patients, but the various states are often conflated or simply confused, or the crucial differences among them are considered unimportant. The sense of nihilism is so pervasive that even the delineation of MCS in the scientific literature has come under attack from some medical quarters. We believe that refining the definitions of brain states is value-neutral, but many physicians have resisted this diagnostic clarification. Some proponents of the right to die have been concerned that this newly identified brain state might erode the hard-won right to forgo life-sustaining therapy. Disability advocates have also voiced their concern, worrying that adding MCS to the categories of impaired brain states could be used nefariously to equate higher-functioning individuals with those in PVS, thus minimizing the value of their lives.

For the record: We support both the right to die and the right to appropriate medical care. We do not see these rights as mutually exclusive, and we view decisions to pursue or refuse care as a matter of ethically balancing the potential benefits and burdens. What is interesting about the discord over MCS is not that the designation could justify either more treatment or less in any particular case, but, rather, how emotionally charged the entire issue has become. As a response to discussion of the designation's scientific basis, the reactions seem way out of proportion. If nothing else, they are a cultural marker for our implicit assumptions about severe cognitive impairment.

The Specter of Psychosurgery

Progress in developing therapies for severe cognitive impairment will also be hampered by the association of deep brain stimulation with psychosurgery and the abuses that have marred its history. News stories about deep brain stimulation often allude to the crude and unjustifiable lobotomies performed by Walter Freeman, a neurologist who performed more than 3,000 procedures on mentally incapacitated patients, and to the specter of mind control. These fears hark back to the debate over psychosurgery in the late 1960s and early 1970s, when Spanish neurophysiologist Jose M. R. Delgado, M.D., advanced the use of an implantable electrode operated by remote control as a way to "psychocivilize society" and cope with the social unrest of the day. Delgado had already won some notoriety by using his "Stimociever" to halt a charging bull in a bullring in 1965. His work entered popular culture through novels and films like Michael Crichton's *Terminal Man* and Stanley Kubrick's *A Clockwork Orange*.

Concerns about the ethics of psychosurgery moved the U.S. Congress to direct the National Commission for the Protection of Human Subjects of Biomedical and Behavioral Research to report on psychosurgery as part of the landmark National Research Act of 1974. The commission did not find that psychosurgery had been used for social control; in fact, it found enough evidence of potential efficacy to recommend that the investigational use of some psychosurgical procedures proceed with appropriate regulation and oversight. This conclusion ran against popular opinion at the time, and popular opinion has never really changed.[6]

Moving Beyond Scientific Stasis

Whatever the merits of deep brain stimulation in neurology and psychiatry (and we believe the merits are potentially great), public perception of its value and even its ethical standing is colored by the legacies of the right-to-die movement and psychosurgery. We believe that this perception does a profound disservice to some of our society's most desperately burdened patients by contributing to their being marginalized and even abandoned. When patients with severe cognitive impairment across the whole spectrum of such conditions are perceived as beyond hope, any potential interventions that might be developed are automatically deemed ethically disproportionate.[7] When this perception is combined with the problem that individuals with severe head trauma often lack the capacity to make decisions and, therefore, cannot give their own consent to enroll in clinical trials, you have a recipe for scientific deadlock.

Future Terry Wallises deserve better. They deserve protection from errors of commission, but equally of omission. They desperately need access to the fruits of science and all the assistance we can provide as they return from the limbo of impaired consciousness and try to reenter the world of human interaction.

The Terry Schiavo Case

THE PLIGHT OF TERRY SCHIAVO, the young woman at the center of the much-publicized legal battle in Florida, illustrates the devastation of the chronic vegetative state following anoxic injury (one that deprives the brain of oxygen). Adults who suffer cardiac arrest and oxygen deprivation to the brain leading to a persistent vegetative state that lasts beyond three months have essentially no statistical chance of further recovery. This grim prognosis rests on a convergence of evidence from studying outcomes in large numbers of patients; in addition, it is supported by evidence of diffuse neuronal damage in the cerebral cortex and other higher brain regions in such an injury. Measurement of patients' brain activity further demonstrates the loss of cerebral function, with the resting metabolic activity in chronic vegetative states following anoxic injuries averaging less than half the level in normal brains. Structural imaging studies in such cases typically reveal widespread neuron loss and cerebral atrophy similar in extent to that observed in the end stages of Alzheimer's disease. Brain electrical activity is grossly disturbed, if evident at all.

Patients in this state may live for years, occasionally smiling, shedding a tear, or briefly appearing to fix their gaze on something. These reflex behaviors in the chronic phase of PVS do not reflect awareness or the potential for further recovery. Many of these facial displays are organized by intrinsic circuits of the brain stem and do not depend on the integrity of higher centers of the brain, including the cortex and thalamus, which are overwhelmingly damaged in patients who chronically remain in a vegetative state. In rare instances, islands of cerebral activity do remain on levels higher than the brain stem, producing fragments of behavior that are not responses to anything in the environment. The presence of these fragmentary behaviors, unfortunately, does not improve the prognosis or suggest greater potential for recovery in patients remaining in vegetative states beyond the three-month period following an anoxic injury. Thus, each patient's examination should be considered in the overall context of the history of his or her illness, along with the results of structural brain imaging and studies of function.

To the untrained observer, the simple appearance of wakefulness is difficult to dissociate from an inference of awareness, especially if this appearance is accompanied by brief, out-of-context, reflexive behaviors that also can be misinterpreted. This emotionally charged situation dictates extraordinarily careful and repeated efforts to reduce the uncertainty in making the diagnosis. In the Schiavo case, many qualified experts have testified that repeated examinations of the patient have revealed a vegetative state, that structural imaging has confirmed the neuron loss and widespread atrophy, and that repeated testing has documented the absence of brain electrical activity. In the aggregate, this evidence is as unequivocal, and lacking in reasons for hope, as any obtainable in these circumstances. —NDS and JJF

CHAPTER 14

A Story of Science, a Story of Grief

Billy's Halo: Love, Science and My Father's Death
by Ruth McKernan
(Joseph Henry Press, 2006; 292 pages, $27.95)

Reviewed by Kevin J. Tracey, M.D.

Kevin J. Tracey, M.D., a neurosurgeon and immunologist, is director and chief executive of the Feinstein Institute for Medical Research and vice president for research at North Shore Long Island Jewish Health Systems. He is the author of *Fatal Sequence: The Killer Within* (Dana Press). Dr. Tracey can be reached at kjtracey@nshs.edu.

BILL MCKERNAN LIVED AN EXTRAORDINARY LIFE
as a British entrepreneur, successful businessman, expert golfer, widely
liked friend, and beloved husband. He was his daughter's hero. In *Billy's
Halo: Love, Science and My Father's Death*, Ruth McKernan, Ph.D., an
accomplished scientist who has spent her career studying the biochemical workings of the nervous system, details the numerous roles she was
forced to assume at the end of her father's life. She was the child of a sick
parent, but also a worried mother, helping her own child cope with what
was happening to a beloved grandparent. She was the communication li-
aison between her family and medical teams that spoke a complex new
language rife with upsetting predictions, and she was a grieving scientist,
grappling with uncertainty about the clinical data stemming from the
disease processes ravaging her dad.

This book is McKernan's chronicle of Billy's illnesses, dissected into
simple, factual prose, not at all unlike what you might find scrawled in
the margins of Ruth's laboratory notebook. Unlike a sterile lab notebook, however, *Billy's Halo* places the science within the context of Billy's life and times through anecdotes that bring both Billy and the science to life. The result is a compelling summary of what Ruth came to
understand about her father's life-threatening infection and leukemia.

Ruth, a neuroscientist at a British pharmaceutical company, remains
first and foremost Billy's daughter, writing to memorialize his achievements and successes. We learn of his roots when, after barely surviving
hospitalization in intensive care for a septic infection, Billy leads his family on a final visit to Blairgowrie, in a farming region of Scotland, where
he lived when he was a boy. Raised a postman's son, Billy was justifiably
proud of the business he founded and nurtured, Molecular Products,
Ltd., a major manufacturer of chemicals that either absorb or produce
breathable gases. The success of his company supported a comfortable
lifestyle for Billy and his family. Nearing retirement, he passed business
operations to his children so that his company would continue to provide for their families.

Ruth writes courageously as she describes her pride in her father's ac-
complishments and her love for him, poignantly noting that his death

crystallized these feelings. She credits Billy for providing the tools she needed to develop her scientific career of experimental design, data analysis, and teaching—the same tools that she applied to exquisite advantage in writing *Billy's Halo.*

Seeking How and Why

A perusal of the table of contents unveils chapter headings that promise a book with treatises on topics of singular complexity, including "Memory," "Consciousness," "Stem Cells," "Genes," "Stress," and "Time." The book lays bare overarching questions in these divergent fields, spanning cell biology, physics, immunology, biochemistry, and neuroscience. Ruth covers it all, because that is where her father's story led her.

Billy's problems started when he developed a bacterial infection in a pimple on his face, just under his left eye. Unfortunately, his immune system's response to that typically trivial problem spun wildly out of control, spilled over into the bloodstream, and put him in a coma. He spent days in the intensive care unit, hovering close to death. Like the rest of her family, Ruth was overwhelmed by the suddenness and magnitude of the bad news that came daily from the doctors and nurses. Billy's problems seemed to go only from bad to worse, when one organ after the next began to fail, as if they had teamed up and decided together to stage a revolt against all things homeostatic.

Writing later, she recalled: "With the benefit of hindsight, and my father's unusual symptoms notwithstanding, everything begins to fit, such that, in the words of Sherlock Holmes, 'a hypothesis gradually becomes a solution.' Once I have done substantial reading, it becomes clear to me that everything my father endured can be accounted for by a bacterial infection. All the symptoms, the trauma, the whole gamut of bodily catastrophes, could be laid at the door of the dishonorable microbe and explained by 'streptococcal toxic shock,' a cascade of events that happens so frequently these days that it has officially become 'a syndrome.'"

This insightful description reveals the duality of this book: it is both Ruth's heart-rending story about grief and loss and the product of her scientific inquiry and reading in search of the underlying cause of her father's disease. Her father has been attacked by microscopic invaders that laid him close to death, and Ruth remains intolerant of partial, dogmatic, unsatisfying, and simplistic explanations. This great man could not possibly have been defeated by a pimple. She enters the fray by reading and writing, and her book reports her answers.

Motivated by her desire to understand Billy's demise and explain it, she brings the science to life, with explanations that arise like a wellspring from her grief. As a scientist, Ruth knows no other way to proceed. She is driven by the need to understand what she knows, what she does not, and what is unknown by all. Her writing is discerning, objective, and accurate, rendering transparent her quest for answers about what happened to her fallen hero.

A Daunting Enemy

The word "syndrome" originates from the Greek word for "a tumultuous concourse running together." As defined by *Stedman's Medical Dictionary,* a syndrome is "the aggregate of symptoms and signs associated with any morbid process, and constituting together the picture of the disease." Billy developed his syndrome of toxic shock from an infected pimple, but I have seen patients who developed toxic shock syndrome from other types of infection, as in a case of complicated childbirth. Early in the disease, the patient with the pimple and the patient with the post-childbirth infection look quite different from each other. Once the signs and symptoms are full-blown, however, the original problem that incited the syndrome may be rendered meaningless, like a scratch on the face of a wounded soldier with a shrapnel injury to his heart and brain. Other problems take center stage.

Indeed, if all of the personal identifiers were magically sponged away or hidden from view, so all that remained was the clinical signs of the "syndrome," then it would be impossible to distinguish one patient from the other as they lay in intensive care. Both would have a bright red rash on the face, fever, shock (dangerously low blood pressure), coma, failure of the kidneys, lungs, and liver, and intolerance of food. These aggregate constellations of seemingly unconnected signs and symptoms form the clinical picture of the syndrome.

Researchers studying the various syndromes caused by infection have begun to uncover proof that unifying mechanisms may explain these apparently unrelated biological failures, and that, surprisingly, the source of the problem lies within the immune system, not with the bacteria. Billy's syndrome was started by infection with a "dishonorable" streptococcus that caused "toxic shock syndrome," but infection with usually innocuous bacteria, such as E. coli, can also initiate similarly devastating syndromes, with names such as "septic shock" and "severe sepsis."

Since the time of Pasteur and his germ theory of disease, scientists have recognized that bacteria can be the underlying cause of illness and that human diseases can be treated by eradicating the invading bacteria. Today's antibiotics are extremely powerful, and the majority of appropriately treated bacterial infections can be cured with these drugs. But sometimes the antibiotics do not prevent the bacteria from aggravating the immune system to the point that it responds by suddenly throwing absolutely all the weapons it possesses at the germs. The uncontrolled attack kills indiscriminately, as bacteria and normal tissues alike become casualties. Even a relatively few bacteria replicating in a pimple or deep inside a bodily recess can switch the immune system into this frenzied, wild attack that damages the body's own organs. Like firing a rocket-propelled grenade to kill a mosquito, weapons are activated in full regardless of whether there are five or five billion bacteria in the affected area of the body. Deadly molecules fill the region, attacking with impartiality, and the damage continues unchecked and unimpeded. Therein was Billy's larger problem.

Once these so named "sepsis syndromes" have evolved to the stage of organ damage, antibiotics do not prevent that damage or improve the chances for survival. Antibiotics can eradicate the inciting germs, but if the bacteria have already activated the immune system attack, antibiotics provide little benefit. A deeply ingrained tendency in medicine is to focus on things that can be readily explained and understood, so doctor-family discussions at the bedside of a patient with a sepsis syndrome inevitably concentrate on the status of the pimple, or the pneumonia, or the urinary tract infection. Often these discussions studiously avoid the immune system's fundamental, causative role in damaging the body. Usually the names of the syndromes are not even spoken; attention instead is rigidly fixed on the body part that may have been originally breached by germs.

Scientific advances in the past twenty years have begun to change this situation, and knowledge is spreading about the identity, function, and structure of molecules that damage the organs and kill the person affected by a sepsis syndrome. This new way of thinking about sepsis, in which the immune system supplies the damaging molecules, has emboldened scientists to develop therapies that neutralize the problem. Knowledge about these molecules and pathways has enabled the development of experimental treatments that can cure severe sepsis syndromes in mice and rats.

For example, it has recently become possible to cure severe sepsis in laboratory animals by administering an antibody that binds to a protein molecule named HMGB1. The goal would be to impair the production of HMGB1, which appears to play a lethal role in sepsis. High levels of this protein are released into the bloodstream during sepsis, and it begins to accumulate in the body's organs. When this occurs, it causes the cells lining the organs to separate from each other and come apart, the result being a leakage of the normally intact barriers within organs that are critical to their function. As the physiological barriers between individual cells fail, water in the blood leaks into the air sacs of the lungs, urinary toxins and bile ooze back into the bloodstream, and bacterial products from the intestines leach into the general circulation. Even worse, as the damage in the organs progresses, the injured cells themselves release more HMGB1, adding to the problem and worsening the state of organ failure. The result is a widespread degeneration of normal physiology that underlies the clinical signs and symptoms of severe sepsis. In the future, it may be possible to prevent the damage caused by HMGB1 using antibodies that have been shown to prevent sepsis syndromes in laboratory animals. These experimental drugs are now being developed for clinical studies.

Despite the abundance of promising studies in animals, the reality is that this knowledge did not benefit Billy and cannot yet benefit other people who face a similar onslaught from a septic infection. Years of additional clinical work are needed before we will know whether HMGB1, or some similar molecule, is the cause of severe sepsis in humans and whether drugs that target its activities can be as effective in preventing organ damage for these people as they have been in laboratory animals.

Science and the Experience of Loss

In Ruth's account, we first meet Billy when he developed sepsis, but we soon learn that some years earlier he had been diagnosed with chronic lymphocytic leukemia, a cancer of the immune system that causes an accumulation of white blood cells. Ruth delves into the pathological underpinnings of this disease, explaining cell birth and cell death and the various ways in which cells can kill or be killed, including a form of cellular suicide termed "apoptosis." Cancer develops if cells either are born too fast or fail to kill themselves on time. The reader is led through fascinating discussions of cellular biochemistry as Ruth seeks to understand the very end of Billy's life.

In the end, Ruth confesses that her dogged pursuit of understanding the processes of death and dying did not ease the pain of her loss: "While Billy lay dying, I sought solace in science. But being a scientist spared me nothing; it provided no shield against fate, no defense against grief. Knowing how emotion affects memory could not keep out the vision of those last few breaths. Knowing how genes and development together could mold personality could not make me love my father more. No matter what we understand of the human mind, it doesn't change what we are."

Though I recommend this book to everyone, I suspect that it will be most sought by students of science and medicine, and by patients, survivors, and families who are dealing with the same illnesses as Billy, sepsis syndromes or chronic lymphocytic leukemia. The inspired writing and heartfelt stories give insight into how science and medicine move forward, often driven by the very personal motivation of the compulsive and curious investigator.

CHAPTER 15

Exploring How Music Works Its Wonders

This Is Your Brain on Music: The Science of a Human Obsession
by Daniel J. Levitin
(New York, Dutton, 2006; $24.95)

Reviewed by David Huron, Ph.D.

David Huron, Ph.D., is a professor of music at Ohio State University, where he directs the Cognitive and Systematic Musicology Laboratory in the School of Music and is also affiliated with the Center for Cognitive Science. His current research focuses on better understanding how music evokes emotion and on cross-cultural comparisons in music perception. He is author of *Sweet Anticipation: Music and the Psychology of Expectation* (MIT Press, 2006). He can be reached at huron.1@osu.edu.

FORTUNE SMILES on those sciences lucky enough to have a popular champion, although not all scientists are enthusiastic about the popular face of their discipline. Professional astronomers found plenty to criticize in the popularizing efforts of Carl Sagan. I know paleobiologists who vehemently disagree with various ideas promoted by Stephen Jay Gould. Some linguists bristle at the mention of Steven Pinker, and many oceanographers roll their eyes when they hear the name Jacques Cousteau.

But whatever the shortcomings of these outstanding popularizers, the benefits that they bring to a research discipline are unquestionable. Popular dissemination of the fruits of research inspires smart people to enter a field, and connecting with colleagues on the other sides of disciplinary fences often leads to important interdisciplinary interaction. Every field of research needs a popularizer.

Strangely, for a subject as inherently popular as music, the science of music has so far failed to produce a writer able to capture the public imagination. For those of us in the discipline, this is a sorry state. When most people hear the phrase "science of music," they tend to think vaguely of right-brain mysticism, music/math associations, or increase-your-brain-power-through-Mozart. But, in fact, a small professional cohort of experimentally oriented researchers, including cognitive and physiological psychologists, neuroscientists, and systematic musicologists, is studying music and musical behaviors. Their research is widely dispersed in dusty journals, but the aggregate findings are genuinely informative about how music works its wonders. Although the story remains incomplete, it is a story worth telling and telling well.

From Rock Musician to Researcher

Making the fruits of music science accessible is the explicit aim of Daniel Levitin's book *This Is Your Brain on Music: The Science of a Human Obsession*. The book's contents unfold against the background of the author's own fascinating personal story. Levitin grew up in the San Francisco area during the heyday of 1960s popular music. A bedroom romance with the guitar matured into a successful career as a rock musician and record producer in Los Angeles. Working with the likes of Steely Dan, Eric Clapton, Stevie Wonder, and k.d. lang, Levitin earned either gold or platinum records for eleven of his projects. Eventually he switched careers, undertook graduate training in cognitive neuroscience, and became a professor of psychology, neuroscience, and music at McGill University in Montreal, where he runs the Levitin Laboratory for Musical Perception, Cognition, and Expertise.

I especially like Levitin's story of how he became transfixed by the acoustic details of recorded sound. His father, eager to gain relief from the racket of Levitin's bedroom stereo, offered to buy his son high-quality headphones if the youngster would turn off the speakers. Levitin took the deal and spent long hours enveloped in the private space of headphone listening—a practice that sensitized him to the arts of stereophonic imaging, sonic balance, and audio quality, all of which would serve him well in the recording industry.

This Is Your Brain on Music has an appealing breezy style. The first two chapters describe how such basic musical sensations as pitch, loudness, timbre, rhythm, and harmony arise. Other chapters cover such topics as emotion, skill acquisition, musical taste, expectation, and categorization. Listening to music is not the passive activity one might suppose; the brain actively anticipates future sounds, even the sounds of unfamiliar music. The brain parses sound patterns into myriad syntactic categories, from scales, chords, and themes to phrases, melodies, and styles.

The book is full of interesting nuggets. We are introduced to compelling infant studies examining the origins of musical preferences, to the structure of jazz improvisations, and to the effects of courtship songs on songbird ovulation. Fun tidbits include a discussion of what happens in the inferior colliculus of a barn owl's brain when the owl hears Strauss's "Blue Danube Waltz," and the fact that much of the funding for the development of magnetic resonance imaging brain scanners came from the profits made by the Beatles' music.

The book is also populated with lots of people. Levitin is a sociable sort and likes to relay his encounters with the famous and the talented. We are treated to personal anecdotes about conversations with Joni

Mitchell, Stevie Wonder, and Francis Crick. We vicariously experience Levitin's Forrest Gump–like presence as we rub elbows with Doc Watson and James Watson.

Perhaps Levitin's unique contribution to research on music lies in his drawing attention to the importance of the cerebellum in music listening. Conventional wisdom suggests that the principal role of the cerebellum relates to coordinating motor movement. But research by Levitin is consistent with studies by Jeremy D. Schmahmann, M.D., and Janet Sherman of Harvard Medical School that point to a much broader role for the cerebellum, including tracking the beat and distinguishing familiar from unfamiliar music. Perhaps most surprising is that the cerebellum plays a role in musically evoked emotions and in the formation and expression of musical taste. These discoveries are consistent with anatomical studies from the 1970s that found direct neural connections between the cerebellum and the hearing organ within the cochlea, which converts sound vibrations into nerve impulses.

An ostinato that recurs throughout the book and comes to the fore in the final chapter is the relationship between music and evolutionary psychology. Archaeological evidence shows that music is very old—much older than a Johnny-come-lately like agriculture. Musical instruments have even been discovered in Neanderthal burial sites, suggesting that music-making may have been characteristic of the entire genus *Homo*. All known cultures (past and present) have engaged in recognizably musical activities, which raises what might be called the $64,000 question in the science of music: Is music a spandrel—a non-adaptive artifact of brain organization, as suggested by Harvard University psychologist Steven Pinker, Ph.D.? Or could music be an evolutionary adaptation in its own right? Levitin, clearly parting company with Pinker, recounts some of the published evidence for music's evolutionary role.

Wanting More

While Levitin informs, and sometimes delights, he also misses some opportunities. *This Is Your Brain on Music* has little to say about the venerable topic of consonance and dissonance, which has been the subject of considerable confusion and misinformation ever since the ancient Greeks speculated about why some sound combinations are more pleasing than others. Yet one of the foremost successes in the science of music has been our understanding of how the physiological properties of the hearing organ influence the perception of dissonance.

Also missing is the pioneering work of psychologist Neil Todd, Ph.D., of Manchester University, on the perception of rhythm and musical "motion." The cochlea of the inner ear contains not only the hearing organ but also the vestibular system, and Todd's work relating the sense of rhythm and motion to auditory-induced activation of the vestibular system is as fascinating as it is unorthodox. A neuroanatomist might well expect that the connections between the cochlea and the cerebellum exist to coordinate movement with vestibular sensations. Given Levitin's interest in the musical role of the cerebellum, a summary of Todd's work would have made a pertinent (and interesting) addition to the book.

Unfortunately, the book is also marred by many small errors, especially with respect to musical acoustics and music theory. For example, an octave is defined as any frequency that is two, three, four (and so on) times some initial frequency; the list should read: two, four, eight, sixteen (and so on). Similarly, the tonic pitch of a scale is constantly referred to as the root, but only chords have roots. It is difficult for the knowledgeable musician to read the text without a little eye-rolling. The book seems to have been written in haste, which is a shame, because Levitin has real gifts as an accessible writer.

A Promising Popularizer

Like religion and sex, music has a special allure that combines mystery with passion. For those of us engaged in music research, this allure is both good and bad—good because it inspires and motivates, bad because it can lead to sloppy thinking. The study of music has always been dominated by music lovers (with potential axes to grind, and sometimes eccentric agendas). People outside the discipline are right to be a little skeptical of the motives of the music-lover/researcher.

The serious student of music, like the sexology researcher, is also haunted by the fear that others may regard the work as frivolous, even suspect. Insecurity is an unfortunate occupational hazard of empirical research in music. The predictable response to the fear of not being taken seriously is to carry out highly technical research and to make certain that the language of scholarly communication is professional, abstruse, and consequently impenetrable.

Music cognition research is in its golden age. Tremendous progress has been made in understanding musically evoked emotions, expectation, memory, the acquisition of musical skills, style, sociocultural factors, and other aspects of this great art. But, regrettably, the best research in our discipline has not been communicated beyond a small coterie of professionals. While Daniel Levitin is not our field's Carl Sagan, *This Is Your Brain on Music* represents a promising start.

CHAPTER 16

The Promise and Perils of "Neural Prostheses"

Shattered Nerves: How Science Is Solving Modern Medicine's Most Perplexing Problem
by Victor D. Chase
(Johns Hopkins University Press, 2006; 289 pages; $27.50)

Reviewed by Edward McKintosh, MRCS

Edward McKintosh, MRCS, spent three years of research in the Department of Neurodegenerative Disease at the Institute of Neurology, London, where he studied Creutzfeldt-Jakob disease, before entering a six-year higher-specialist neurosurgery training program at King's College Hospital. He can be reached at edward1812@yahoo.com.

IN *SHATTERED NERVES,* Victor D. Chase explores the development of "neural prostheses," devices that interact directly with nerves or the brain to try to compensate for the lost functions of an injured nervous system. As he himself says, a comprehensive record of this field would require something more the size of an encyclopedia than the 289 pages he has produced. Instead, Chase furnishes an overview of the field, its technology, and the people in it. He achieves his formidable aim, enabling the reader to make connections between scientific endeavor and its application to the lives of the vivid individuals to whom he introduces us.

Any injury affecting the ability to use a limb has consequences beyond the purely physical. The injury, the reactions of other people, and the altered sense of self that patients experience can all have a profound impact. Generally, spinal injuries hinder not just the capacity to use limbs but also the ability to control bladder and bowel function and to have sex. These injuries, therefore, not only affect a person's ability to carry out day-to-day tasks but may also make reliable control of continence impossible, leading to yet greater prejudice and social isolation. Similar problems plague those with injuries to other parts of the nervous system, such as vision and hearing.

Chase engages his readers by encouraging them to follow the patients' and researchers' learning process as they seek to develop and use neural prostheses. By revealing what the science means to a specific person—a quadriplegic woman who becomes able to walk down the aisle at her wedding, for instance—he makes the scientific knowledge accessible and gives resonance to its significance.

Both the general subject matter and the individual stories could have made for a long, sad read, but Chase uses the stories well, showing the huge amounts of determination that people are able to generate to overcome their injuries and also the benevolence and fortitude that they are capable of displaying by acting as experimental subjects.

From Electricity to Ethics

Most human tissues have some ability to re-grow after injury or surgery. The brain, spine, eyes, and ears generally do not. Injuries to these organs are permanent, creating particular problems for the surgeons and physicians who treat injuries, tumors, or other disorders affecting the nervous system. Scientists now know that nerve cells can regenerate in certain circumstances, and considerable research effort is devoted to this field. So far, however, no major surgical or pharmacological solutions are available for day-to-day clinical use, hence the need for neural prostheses.

I have often felt that the rapid increase in the complexity and capability of consumer electronic devices, as well as their continuing miniaturization, should offer hope for medical applications. If mass-market cameras are now sophisticated enough to focus on individual faces, produce images that can be printed at poster size, and still fit in the palm of one's hand, we might speculate that an eyeball-size camera is not far away. Chase describes the advances that this technological sophistication has allowed and the problems that remain to be conquered. He starts by explaining how nerve and muscle cells use electricity to function and outlining the basic organization of the central and peripheral nervous systems. This discussion provides the basis for later parts of the book, in which he describes research of the eighteenth, nineteenth, and twentieth centuries, including experiments in how to generate electricity in the first place and some of the first experiments on the nervous system, such as the results of placing an electrode in one's ear.

Chase then takes us through the problems of designing, coordinating, controlling, and powering neural prostheses, as well as the postoperative tuning necessary to enable an exceedingly complex brain to derive maximum benefit from a device with a very limited number of input channels. He describes the experiments that help researchers determine the physiological rules governing neuronal and muscular stimulation and also the technology necessary to develop devices that can reliably stimulate nerves and last many years in the body without producing toxicity. He explains the chemical challenges involved in selecting the correct materials for making electrodes and the engineering challenges in manufacturing these electrodes at microscopic sizes. Finally, he delves into the ethics of neural implantation, such as whether we should implant prostheses in non-injured humans in an attempt to augment their function.

Seeking Hope and Healing

Chase provides a very thorough overview of this field, from A.D. 50 to the present day and from prostheses designed to help hearing, to those that enable people to use paralyzed hands and receive enough feedback to control them. He provides insights into the lives of patients who need implants and the emotional and physical upheavals they go through, as well as showing very clearly the successes that research has had to date and the profound challenges that remain. Most important, by focusing on the stories of a colorful cast of patients and researchers, he makes *Shattered Nerves* easy and entertaining to read. Chase's powers of description and his insights into character are put to good use throughout the book, and he clearly has a great deal of admiration for the individuals he portrays. In a society that too often defines people by their injuries, Chase's patients speak for themselves and come to define themselves instead by their aspirations.

The case studies in *Shattered Nerves* illustrate how relatively small improvements (for example, to perceive a few dots of light or channels of sound) may not only produce surprising increases in a patient's ability to function physically but also have dramatic psychological benefits. One patient explains how his partial and then complete loss of hearing forced him to reevaluate the nature of remedy: "Our concept of healing and recovery often takes on new meaning, one that is focused on the spiritual and psychological, rather than a physical recovery. If we can no longer expect to join nerves back together, we can instead try to reconfigure the soul and the self." In this context, the reader can understand the disproportionate feelings of emotional well-being experienced by patients who make seemingly modest gains.

One of the problems facing researchers and clinicians involved in treating injuries to the brain and nervous system is that both patients and their relatives are understandably keen to seize upon any chance of improvement. Hope and the determination to overcome problems are vital attributes for both patients and researchers, but this technology is still in its infancy and false hopes can be damaging. One of the many strengths of *Shattered Nerves* is that through the author's description of the numerous difficulties that have been overcome in reaching the current level of implant sophistication, readers gain a clear understanding of the difficulties that must yet be overcome in order to make further advances.

A Broadly Appealing Story

Perhaps when Chase introduces us to "the grandfather of neural prostheses," Giles Brindley, and tells the hilarious and slightly disturbing story of this "grandfather" dropping his trousers in front of a lecture theater audience, one should be grateful that there are no illustrations in the book. Chase rarely uses scientific jargon and explains it well when he does. He is very good at describing the prostheses and procedures, and the absence of illustrations is more than compensated for by the amount of information that he covers and his clear writing style. On balance, however, the book would benefit from diagrams of the parts of the nervous system affected by various conditions and photographs or illustrations of the prostheses and surgical procedures described.

One of Chase's most notable achievements in *Shattered Nerves* is that he has written about his narrow, specialized subject matter with such broad appeal. The reader already familiar with this field is still likely to end up better informed and certainly entertained by Chase's character sketches, while readers from a non-neuroscience background will find an extremely thorough and balanced introduction to neural prostheses, presented in an enjoyable and accessible way. Irrespective of one's profession or background, Chase's approach to his patients and to his subject must surely inspire greater endeavor in us all.

Endnotes

I. STROKE: WE CAN AND MUST DO BETTER
Improving Stroke Prevention and Treatment Now

1. Gorelick PB. Stroke prevention. *Archives of Neurology.* 1995;52:347-355.

2. Gorelick PB. Stroke prevention therapy beyond antithrombotics: unifying mechanisms in ischemic stroke pathogenesis and implications for therapy: an invited review. *Stroke.* 2002;33:862-875.

3. Chobanian AV, Bakris GL, Black HR, et al; National Heart, Lung, and Blood Institute Joint National Committee on Prevention, Detection, Evaluation, and Treatment of High Blood Pressure; National High Blood Pressure Education Program Coordinating Committee. The Seventh Report of the Joint National Committee on Prevention, Detection, Evaluation, and Treatment of High Blood Pressure: the JNC 7 report. *Journal of the American Medical Association.* 2003;289:2560-2572.

4. Adams HP, Brott TG, Furlan AJ, et al. Use of thrombolytic drugs. A supplement to the guidelines for the management of patients with acute ischemic stroke. A statement for Health Care Professionals from a special writing group of the Stroke Council American Heart Association. *Stroke.* 1996;27:1711-1718.

5. Quality Standards Subcommittee of the American Academy of Neurology, Practice advisory: thrombolytic therapy for acute ischemic stroke—summary statement. *Neurology.* 1996;47:835-839.

Searching for a New Strategy to Protect the Brain

1. Fisher M, Ratan R. New perspectives on developing acute stroke therapy. *Annals of Neurology.* 2003; 53:10-20.

2. Gladstone DJ, Black SE, Hakim AM. Toward wisdom from failure: lessons from neuroprotective stroke trials and new therapeutic directions. *Stroke.* 2002; 33:2123-2136.

3. Grotta J. Neuroprotection is unlikely to be effective in humans using current trial designs. *Stroke.* 2002; 33:306-307.

4. Bazan NG: Neuroprotectin D1 (NPD1): A DHA-derived mediator that protects brain and retina against cell injury-induced oxidative stress. *Brain Pathology* 15:159-166, 2005.

5. Mukherjee PK, Marcheselli VL, Serhan CN, Bazan NG. Neuroprotectin D1: a docosahexaenoic acid-derived docosatriene protects human retinal pigment epithelial cells from oxidative stress. *Proceedings of the National Academy of Science USA.* 2004; 101:8491-8496.

6. Marcheselli V.L., Hong S., Lukiw W.J., Tian X.H., Gronert K., Musto A., Hardy M., Gimenez J.M., Chiang N., Serhan C.N., and Bazan N.G. (2003) Novel docosanoids inhibit brain ischemia-reperfusion-mediated leukocyte infiltration and proinflammatory gene expression. *Journal of Biological Chemistry.* 278, 43807-43817.

7. Belayev L, Marcheselli Vl, Khoutorova L, Rodriguez de Turco EB, Busto R, Ginsberg MD, Bazan NG: Docosahexaenoic acid complexed to albumin elicits high-grade ischemic neuroprotection. *Stroke* 36:118-123, 2005.

8. Belayev L, Liu Y, Zhao W, Busto R, Ginsberg MD. Human albumin therapy of acute ischemic stroke: marked neuroprotective efficacy at moderate doses and with a broad therapeutic window. *Stroke.* 2001; 32:553-560.

6. KNOWING SIN: MAKING SURE GOOD SCIENCE DOESN'T GO BAD

1. J. Robert Oppenheimer, "Physics in the Contemporary World," Arthur Dehon Little memorial lecture at the Massachusetts Institute of Technology, November 25, 1947 (Cambridge, MA, 1947). Oppenheimer's lecture, with this quotation, was also published in 4 Bulletin Atomic Scientists 65, 66 (March 1948). He used the same language in an interview with Time magazine: "Expiation," *Time*, p. 94 (February 23, 1948).

2. Michael S. Gazzaniga, *The Ethical Brain* (Dana Press, 2005)

3. Judy Illes, ed., Neuroethics: Defining the Issues in Theory, Practice and Policy (Oxford University Press, 2006)

8. BRINGING THE BRAIN OF THE CHILD WITH AUTISM BACK ON TRACK

1.Casanova MF, Buxhoeveden D, Gomez J. (2003). Disruption in the inhibitory architecture of the cell minicolumn: implications for autism. *Neuroscientist* 9, 496-507.

2. Bauman ML, Kemper TL (1994). Neuroanatomic observations of the brain in autism. In: Bauman ML, Kemper TL (eds.) *The Neurobiology of Autism* Johns Hopkins Press, Baltimore, pp. 119 141.

3. Lam KS, Aman MG, Arnold LE. Neurochemical correlates of autistic disorder: a review of the literature. *Research in Developmental Disabilities.* 2006 May-Jun;27(3):254-89.

4. Chugani DC, Muzik O, Behen ME, Rothermel RD, Lee J, Chugani HT. Developmental changes in brain serotonin synthesis capacity in autistic and non-autistic children. *Annals of Neurology* 1999; 45: 287-295.

5. Bennett-Clarke CA, Chiaia NL, Rhoades RW (1996). Thalamocortical afferents in rat transiently express high-affinity serotonin uptake sites. *Brain Research* 733, 301-306.

6. Bennett-Clarke CA, Leslie MJ, Lane RD, Rhoades RW (1994). Effect of serotonin depletion on vibrissae-related patterns in the rat's somatosensory cortex. *Journal of Neuroscience* 14, 7594-7607.

7. Gaspar P, Cases O, Maroteaux L (2003). The developmental role of serotonin: news from mouse molecular genetics. *Nature Reviews Neuroscience* 4, 1002-1012.

8. Gross C, Zhuang X, Stark K, Ramboz S, Oosting R, Kirby L, Santarelli L, Beck S, Hen R (2002). Serotonin1A receptor acts during development to establish normal anxiety-like behaviour in the adult. *Nature* 416, 396-400.

9. Andersen SL (2003). Trajectories of brain development: point of vulnerability or window of opportunity? *Neuroscience and Biobehavioral Reviews* 27, 3-18.

10. Huttenlocher PR (1979). Synaptic density in human frontal cortex—developmental changes and effects of aging. *Brain Research* 163, 195-205.

11. Akers KG, Nakazawa M, Romeo RD, Connor JA, McEwen BS, Tang AC. Early life modulators and predictors of adult synaptic plasticity. *European Journal of Neuroscience* 2006 Jul;24(2):547-554.

12. Goodman C, Shatz C (1993). Developmental mechanisms that generate precise patterns of neuronal connectivity. *Cell* 72 Suppl 10, 77-98.

13. Edagawa Y, Saito H, Abe K (2001). Endogenous serotonin contributes to a developmental decrease in long-term potentiation in the rat visual cortex. *Journal of Neuroscience* 21, 1532-1537.

9. TOWARD A NEW TREATMENT FOR TRAUMATIC MEMORIES

1. Pitman, RK, Sanders, KM, Zusman, RM, et al. Pilot study of secondary prevention of posttraumatic stress disorder with propranolol. *Biological Psychiatry* Jan. 15 2002; 51(2): 189–192.

2. Lewis, DJ. Psychobiology of active and inactive Memory. *Psychological Bulletin* 1979; 86(5): 1054–1083.

3. Dębiec, J, and LeDoux, JE. Disruption of reconsolidation but not consolidation of auditory fear conditioning by noradrenergic blockade in the amygdala. *Neuroscience* 2004; 129(2): 267–272.

4. Miller, MM, Altemus, M, Dębiec, J, LeDoux, JE, and Phelps, EA. Propranolol impairs reconsolidation of conditioned fear in humans. Program No. 208.2. *2004 Abstract Viewer/Itinerary Planner.* Washington, DC: Society for Neuroscience, 2004. Online.

5. Davis, M, Ressler, K, Rothbaum, BO, Richardson, R. Effects of D-cycloserine on extinction: translation from preclinical to clinical work. *Biological Psychiatry* 2006; 60(4): 369–375.

6. Hofmann, SG, Meuret, AE, Smits, JAJ, et al. Augmentation of exposure therapy with d-cycloserine for social anxiety disorder. *Archives of General Psychiatry* 2006; 63: 298–304.

7. Kass, LR (Ed.) *Beyond Therapy: Biotechnology and the Pursuit of Happiness.* A Report of the President's Council on Bioethics. New York. Dana Press, 2005.

11. TRANSFORMING DRUG DEVELOPMENT THROUGH BRAIN IMAGING

1. Alexander GE, Chen K, Pietrini P, Rapoport SI, and Reiman EM. Longitudinal PET Evaluation of Cerebral Metabolic Decline in Dementia: A Potential Outcome Measure in Alzheimer's Disease Treatment Studies. *American Journal of Psychiatry* 2002; 159(5): 738-745.

2. Reiman EM, Chen K, Alexander GE, Caselli RJ, Bandy D, Osborne D, Saunders AM, and Hardy J. Correlations Between Apolipoprotein Epsilon4 Gene Dose and Brain-Imaging Measurements of Regional Hypometabolism. *Proceedings of the National Academy of Science* 2005; 102(23): 8299-8302.

3. Schott JM, Price SL, Frost C, Whitwell JL, Rossor MN, and Fox NC. Measuring

Atrophy in Alzheimer Disease: A Serial MRI Study Over 6 and 12 Months. *Neurology* 2005; 65(1): 119-124.

4. Matthews, PM, Honey, GD, and Bullmore, ET. Applications of fMRI in Translational Medicine and Clinical Practice. *Nature Reviews Neuroscience* 2006; 7: 733-744.

5. Mattay VS, Goldberg TE, Fera F, Hariri AR, Tessitore A, Egan MF, Kolachana B, Callicott JH, and Weinberger DR. Catechol O-methyltransferase val158-met Genotype and Individual Variation in the Brain Response to Amphetamine. *Proceedings of the National Academy of Science* 2003; 100(10): 6186-6191.

6. Heinz A, Siessmeier T, Wrase J, Hermann D, Klein S, Grusser SM, Flor H, Braus DF, Buchholz HG, Grunder G, Schreckenberger M, Smolka MN, Rosch F, Mann K, and Bartenstein P. Correlation Between Dopamine D(2) Receptors in the Ventral Striatum and Central Processing of Alcohol Cues and Craving. *American Journal of Psychiatry* 2004; 161(10): 1783-1789.

7. Paulus MP, Feinstein JS, Castillo G, Simmons AN, and Stein MB. Dose-Dependent Decrease of Activation in Bilateral Amygdala and Insula by Lorazepam During Emotion Processing. *Archives of General Psychiatry* 2005; 62(3): 282-288.

8. Ploghaus A, Tracey I, Gati JS, Clare S, Menon RS, Matthews PM, and Rawlins JN. Dissociating Pain from Its Anticipation in the Human Brain. *Science* 1999; 284(5422): 1979-1981.

9. Wise RG, Rogers R, Painter D, Bantick S, Ploghaus A, Williams P, Rapeport G, and Tracey I. Combining fMRI with a Pharmacokinetic Model to Determine Which Brain Areas Activated by Painful Stimulation Are Specifically Modulated by Remifentanil. *Neuroimage* 2002; 16(4): 999-1014.

10. Petrovic P, Dietrich T, Fransson P, Andersson J, Carlsson K, and Ingvar M. Placebo in Emotional Processing–Induced Expectations of Anxiety Relief Activate a Generalized Modulatory Network. *Neuron* 2005; 46(6): 957-969.

11. Trip SA, Schlottmann PG, Jones SJ, Li WY, Garway-Heath DF, Thompson AJ, Plant GT, and Miller DH. Optic Nerve Atrophy and Retinal Nerve Fibre Layer Thinning Following Optic Neuritis: Evidence That Axonal Loss Is a Substrate of MRI-Detected Atrophy. *Neuroimage* 2006; 31(1): 286-293.

The Long, Sometimes Bumpy Road of Drug Development

1. Schenk D, Games D, and Seubert P. Potential Treatment Opportunities for Alzheimer's Disease Through Inhibition of Secretases and Abeta Immunization. *Journal of Molecular Neuroscience* 2001; 17(2): 259-267.

2. Thal LJ, Kantarci K, Reiman EM, Klunk WE, Weiner MW, Zetterberg H, Galasko D, Pratico D, Griffin S, Schenk D, and Siemers E. The Role of Biomarkers in Clinical Trials for Alzheimer Disease. *Alzheimer Disease and Associated Disorders* 2006; 20(1): 6-15.

12. HARDWIRED FOR HAPPINESS

1. Olds, J, and Milner, P. Positive Reinforcement Produced by Electrical Stimulation of the Septal Area and Other Regions of the Rat Brain. *Journal of Comparative and Physiological Psychology* 1954; 47: 419–428.

2. Berridge, KC. Motivation Concepts in Behavioral Neuroscience. *Physiology and Behavior* 2004; 81(2): 179–209.

3. Pert, C. *Molecules of Emotion: The Science Behind Mind-Body Medicine.* New York. Scribner, 1999.

4. Damasio, AR, Grabowski, TJ, Bechara, A, Damasio, H, Ponto, LL, Parvizi, J, and Hichwa, RD. Subcortical and Cortical Brain Activity During the Feeling of Self-Generated Emotions. *Nature Neuroscience* 2000; 3: 1049–1056.

5. Damasio, A. Feeling of What Happens: Body, Emotion, and the Making of Consciousness. London. Heinemann, 1999.

6. Levesque, J Eugene F, Joanette, Y, Paquette, V, Mensour, B, Beaudoin, G, Leroux, J-M, Bourgouin, P, and Beauregard, M. Neural Circuitry Underlying Voluntary Suppression of Sadness. *Biological Psychiatry* 2003; 53: 502–510.

7. Murphy, FC, Nimmo-Smith, I, and Lawrence, AD. Functional Neuroanatomy of Emotions: A Meta-analysis. *Cognitive, Affective, and Behavioral Neuroscience* 2003; 3(3): 207–233.

8. Davidson, RJ. Toward a Biology of Personality and Emotion. *Annals of the New York Academy of Sciences* 2001; 935: 191–207.

9. Ekman, P. Basic Emotions. In T Dalgleish and MJ Power (Eds.), *Handbook of Cognition and Emotion.* Chichester, UK. Wiley, 1999: 45–60.

10. Panksepp, J. Emotions as Natural Kinds Within the Mammalian Brain. In M Lewis and JM Haviland-Jones (Eds.), *Handbook of Emotions.* New York. Guilford, 2000: 2nd ed., 137–156.

11. Lykken, D. *Happiness: The Nature and Nurture of Joy and Contentment.* New York. St. Martin's Griffin, 2000.

12. Swanson, LW. *Cerebral Hemisphere Regulation of Motivated Behavior* (1). Brain Research 2000; 886: 113–164.

13. HOPE FOR "COMATOSE" PATIENTS

1. Giacino, JT, Ashwal, S, Childs, N, et al. "The minimally conscious state: definition and diagnostic criteria." *Neurology* 2002; 58(3): 349-353.

2. Jennett, B. The vegetative state: medical facts, ethical and legal dilemmas. Cambridge University Press, 2002. and Jennett, B and Plum, F. "Persistent vegetative state after brain damage. A syndrome in search of a name." *Lancet* 1972; 1: 734-737.

3. Schiff, ND, Ribary, U, Moreno, DR, et al. "Residual Cerebral Activity and Behavioural Fragments Can Remain in the Persistently Vegetative Brain." *Brain* 2002; 125: 1210-1234.

4. Laureys, S, Faymonville, ME, Peigneux, P, et al. "Cortical processing of noxious somatosensory stimuli in the persistent vegetative state." *Neuroimage* 2002; 17(2): 732-741.

5. Hirsch, J, Kamal A, Moreno, D, et al. "fMRI reveals intact cognitive systems for two minimally conscious patients." *Society for Neuroscience, Abstracts* 2001; 271(1):1397.

6. Fins, JJ. "From Psychosurgery to Neuromodulation and Palliation: History's Lessons for the Ethical Conduct and Regulation of Neuropsychiatric Research." *Neurosurgery Clinics of North America* 2003; 14(2): 303-319.

7. Fins, JJ. "Constructing an Ethical Stereotaxy for Severe Brain Injury: Balancing Risks, Benefits and Access." *Nature Reviews Neuroscience* 2003; 4: 323-327.

Index

Other Dana Press
Books and Periodicals

www.dana.org/books/press

Books For General Readers

Brain and Mind:

MIND WARS: Brain Research and National Defense
Jonathan Moreno, Ph.D.

A leading ethicist examines national security agencies' work on defense applications of brain science, and the ethical issues to consider.

Cloth 210 pp. 1-932594-16-7 • $23.95

THE DANA GUIDE TO BRAIN HEALTH: A Practical Family Reference from Medical Experts (with CD-ROM)
Floyd E. Bloom, M.D., M. Flint Beal, M.D., and David J. Kupfer, M.D., Editors
Foreword by William Safire

The only complete, authoritative family-friendly guide to the brain's development, health, and disorders. *The Dana Guide to Brain Health* offers ready reference to our latest understanding of brain diseases as well as information to help you participate in your family's care. 16 full-color illustrations and more than 200 black-and-white drawings.

Paper (with CD-ROM) 733 pp. 1-932594-10-8 • $25.00

THE CREATING BRAIN: The Neuroscience of Genius
Nancy C. Andreasen, M.D., Ph.D.

A noted psychiatrist and bestselling author explores how the brain achieves creative breakthroughs, including questions such as how creative people are different and the difference between genius and intelligence. She also describes how to develop our creative capacity. 33 illustrations/photos.

Cloth 197 pp. 1-932594-07-8 • $23.95

THE ETHICAL BRAIN
Michael S. Gazzaniga, Ph.D.

Explores how the lessons of neuroscience help resolve today's ethical dilemmas, ranging from when life begins to free will and criminal responsibility. The author, a pioneer in cognitive neuroscience, is a member of the President's Council on Bioethics.

Cloth 201 pp.1-932594-01-9 • $25.00

A GOOD START IN LIFE: Understanding Your Child's Brain and Behavior from Birth to Age 6
Norbert Herschkowitz, M.D., and Elinore Chapman Herschkowitz

The authors show how brain development shapes a child's personality and behavior, discussing appropriate rule-setting, the child's moral sense, temperament, language, playing, aggression, impulse control, and empathy. 13 illustrations.

Cloth 283 pp. 0-309-07639-0 • $22.95
Paper (Updated with new material) 312 pp. 0-9723830-5-0 • $13.95

BACK FROM THE BRINK: How Crises Spur Doctors to New Discoveries about the Brain
Edward J. Sylvester

In two academic medical centers, Columbia's New York Presbyterian and Johns Hopkins Medical Institutions, a new breed of doctor, the neurointensivist, saves patients with life-threatening brain injuries. 16 illustrations/photos.

Cloth 296 pp. 0-9723830-4-2 • $25.00

THE BARD ON THE BRAIN: Understanding the Mind Through the Art of Shakespeare and the Science of Brain Imaging
Paul Matthews, M.D., and Jeffrey McQuain, Ph.D. Foreword by Diane Ackerman

Explores the beauty and mystery of the human mind and the workings of the brain, following the path the Bard pointed out in 35 of the most famous speeches from his plays. 100 illustrations.

Cloth 248 pp. 0-9723830-2-6 • $35.00

STRIKING BACK AT STROKE: A Doctor-Patient Journal
Cleo Hutton and Louis R. Caplan, M.D.

A personal account with medical guidance from a leading neurologist for anyone enduring the changes that a stroke can bring to a life, a family, and a sense of self. 15 illustrations.

Cloth 240 pp. 0-9723830-1-8 • $27.00

UNDERSTANDING DEPRESSION: What We Know and
What You Can Do About It

J. Raymond DePaulo Jr., M.D., and Leslie Alan Horvitz.

Foreword by Kay Redfield Jamison, Ph.D.

What depression is, who gets it and why, what happens in the brain, troubles that come
with the illness, and the treatments that work.

Cloth 304 pp. 0-471-39552-8 • $24.95
Paper 296 pp. 0-471-43030-7 • $14.95

KEEP YOUR BRAIN YOUNG: The Complete Guide to Physical and
Emotional Health and Longevity

Guy McKhann, M.D., and Marilyn Albert, Ph.D.

Every aspect of aging and the brain: changes in memory, nutrition, mood, sleep, and sex,
as well as the later problems in alcohol use, vision, hearing, movement, and balance.

Cloth 304 pp. 0-471-40792-5 • $24.95
Paper 304 pp. 0-471-43028-5 • $15.95

THE END OF STRESS AS WE KNOW IT

Bruce McEwen, Ph.D., with Elizabeth Norton Lasley

Foreword by Robert Sapolsky

How brain and body work under stress and how it is possible to avoid its debilitating
effects.

Cloth 239 pp. 0-309-07640-4 • $27.95
Paper 262 pp. 0-309-09121-7 • $19.95

IN SEARCH OF THE LOST CORD: Solving the Mystery of
Spinal Cord Regeneration

Luba Vikhanski

The story of the scientists and science involved in the international scientific race to find
ways to repair the damaged spinal cord and restore movement. 21 photos; 12 illustrations.

Cloth 269 pp. 0-309-07437-1 • $27.95

THE SECRET LIFE OF THE BRAIN

Richard Restak, M.D.

Foreword by David Grubin

Companion book to the PBS series of the same name, exploring recent discoveries about
the brain from infancy through old age.

Cloth 201 pp. 0-309-07435-5 • $35.00

THE LONGEVITY STRATEGY: How to Live to 100 Using the
Brain-Body Connection

David Mahoney and Richard Restak, M.D.

Foreword by William Safire

Advice on the brain and aging well.

Cloth 250 pp. 0-471-24867-3 • $22.95

Paper 272 pp. 0-471-32794-8 • $14.95

STATES OF MIND: New Discoveries about How Our Brains Make Us
Who We Are

Roberta Conlan, Editor

Adapted from the Dana/Smithsonian Associates lecture series by eight of the country's
top brain scientists, including the 2000 Nobel laureate in medicine, Eric Kandel.

Cloth 214 pp. 0-471-29963-4 • $24.95

Paper 224 pp. 0-471-39973-6 • $18.95

The Dana Foundation Series On Neuroethics:

HARD SCIENCE, HARD CHOICES: Facts, Ethics, and Policies
Guiding Brain Science Today

Sandra J. Ackerman, Editor

Top scholars and scientists discuss new and complex medical and social ethics brought
about by advances in neuroscience. Based on an invitational meeting co-sponsored by
the Library of Congress, the National Institutes of Health, the Columbia University
Center for Bioethics, and the Dana Foundation.

Paper 152 pp. 1-932594-02-7 • $12.95

NEUROSCIENCE AND THE LAW: Brain, Mind, and the
Scales of Justice

*Brent Garland, Editor. With commissioned papers by Michael S. Gazzaniga,
Ph.D., and Megan S. Steven; Laurence R. Tancredi, M.D., J.D.; Henry T.
Greely, J.D.; and Stephen J. Morse, J.D., Ph.D.*

How discoveries in neuroscience influence criminal and civil justice, based on an invitational meeting of 26 top neuroscientists, legal scholars, attorneys, and state and federal
judges convened by the Dana Foundation and the American Association for the Ad-
vancement of Science.

Paper 226 pp.1-932594-04-3 • $8.95

BEYOND THERAPY: Biotechnology and the Pursuit of Happiness.
A Report of the President's Council on Bioethics

Special Foreword by Leon R. Kass, M.D., Chairman.

Introduction by William Safire

Can biotechnology satisfy human desires for better children, superior performance, age-less bodies, and happy souls? This report says these possibilities present us with profound ethical challenges and choices. Includes dissenting commentary by scientist members of the Council.

Paper 376 pp. 1-932594-05-1 • $10.95

NEUROETHICS: Mapping the Field. Conference Proceedings.

Steven J. Marcus, Editor

Proceedings of the landmark 2002 conference organized by Stanford University and the University of California, San Francisco, at which more than 150 neuroscientists, bio-ethicists, psychiatrists and psychologists, philosophers, and professors of law and public policy debated the ethical implications of neuroscience research findings. 50 illustrations.

Paper 367 pp. 0-9723830-0-x • $10.95

Immunology:

RESISTANCE: The Human Struggle Against Infection

Norbert Gualde, M.D., translated by Steven Rendall

Traces the histories of epidemics and the emergence or re-emergence of diseases, illustrating how new global strategies and research of the body's own weapons of immunity can work together to fight tomorrow's inevitable infectious outbreaks.

Cloth 219 pp. 1-932594-00-0 $25.00

FATAL SEQUENCE: The Killer Within

Kevin J. Tracey, M.D.

An easily understood account of the spiral of sepsis, a sometimes fatal crisis that most often affects patients fighting off nonfatal illnesses or injury. Tracey puts the scientific and medical story of sepsis in the context of his battle to save a burned baby, a sensitive telling of cutting-edge science.

Cloth 225 pp. 1-932594-06-x • $23.95
Paper 225 pp. 1-932594-09-4 • $12.95

Arts Education:

A WELL-TEMPERED MIND: Using Music to Help Children
Listen and Learn

Peter Perret and Janet Fox

Foreword by Maya Angelou

Five musicians enter elementary school classrooms, helping children learn about music and contributing to both higher enthusiasm and improved academic performance. This charming story gives us a taste of things to come in one of the newest areas of brain research: the effect of music on the brain. 12 illustrations.

Cloth 231 pp. 1-932594-03-5 • $22.95
Paper 231 pp. 1-932594-08-6 • $12.00

Free Educational Books

(Information about ordering and downloadable PDFs are available at www.dana.org.)

PARTNERING ARTS EDUCATION: A Working Model from ArtsConnection

This publication describes how classroom teachers and artists learned to form partnerships as they built successful residencies in schools. *Partnering Arts Education* provides insight and concrete steps in the ArtsConnection model. 55 pp.

ACTS OF ACHIEVEMENT: The Role of Performing Arts Centers in Education.

Profiles of more than 60 programs, plus eight extended case studies, from urban and rural communities across the United States, illustrating different approaches to performing arts education programs in school settings. Black-and-white photos throughout. 164 pp.

PLANNING AN ARTS-CENTERED SCHOOL: A Handbook

A practical guide for those interested in creating, maintaining, or upgrading arts-centered schools. Includes curriculum and development, governance, funding, assessment, and community participation. Black-and-white photos throughout. 164 pp.

THE DANA SOURCEBOOK OF BRAIN SCIENCE: Resources for Teachers and Students, Fourth Edition

A basic introduction to brain science, its history, current understanding of the brain, new developments, and future directions. 16 color photos; 29 black-and-white photos; 26 black-and-white illustrations. 160 pp.

THE DANA SOURCEBOOK OF IMMUNOLOGY: Resources for Secondary and Post-Secondary Teachers and Students

An introduction to how the immune system protects us, what happens when it breaks down, the diseases that threaten it, and the unique relationship between the immune system and the brain. 5 color photos; 36 black-and-white photos; 11 black-and-white illustrations. 116 pp. ISSN: 1558-6758

PERIODICALS

Dana Press also offers several periodicals dealing with arts education, immunology, and brain science. These periodicals are available free to subscribers by mail. Please visit www .dana.org.